Technological Innovation and Prize Incentives

To Mercedes

Technological Innovation and Prize Incentives

The Google Lunar X Prize and Other Aerospace Competitions

Luciano Kay

Center for Nanotechnology in Society, University of California, Santa Barbara, USA and Georgia Tech Program in Science, Technology and Innovation Policy, Georgia Institute of Technology, Atlanta, USA

Edward Elgar
Cheltenham, UK • Northampton, MA, USA

Published by
Edward Elgar Publishing Limited
The Lypiatts
15 Lansdown Road
Cheltenham
Glos GL50 2JA
UK

Edward Elgar Publishing, Inc.
William Pratt House
9 Dewey Court
Northampton
Massachusetts 01060
USA

A catalogue record for this book
is available from the British Library

Library of Congress Control Number: 2012944474

ISBN 978 1 78100 647 4

Typeset by Servis Filmsetting Ltd, Stockport, Cheshire
Printed and bound by MPG Books Group, UK

Contents

Figures

Tables

Foreword

Offering monetary inducement prizes to stimulate and accelerate the achievement of scientific, engineering and technological goals has a celebrated lineage. In 1714, the British Government established the Longitude Prize to encourage the development of a feasible way to measure a ship's east-west position while at sea. Over the subsequent half-century, multiple recipients earned cash awards for inventions and methods that helped to solve the problem of determining longitude. In 1919, the $25000 Orteig Prize was announced for the first nonstop flight between New York and Paris. Advances in aviation technology aided Charles Lindbergh to win this prize in 1927. More recently, in 2004, Scaled Composites earned the $10 million Ansari X Prize – set up in 1996 to reward the first non-governmental team to launch, and then re-launch within two weeks, a reusable space vehicle. The Ansari X Prize not only motivated the private exploration of space, with numerous teams worldwide vying to reach the goal, but also spurred renewed activity in using inducement prizes to address technological, medical, environmental, and other societal challenges. Today, in a growing number of countries, private foundations, government agencies, research organizations, and private companies are offering inducement prizes, ranging in value from the equivalent of a few thousand to many millions of dollars. Advocates suggest that inducement prizes accelerate innovation by promoting excitement, competition and entrepreneurship and by leveraging resources that typically greatly exceed the value of the prize itself. On the face of it, this is a compelling argument. Yet, notwithstanding the great increase of investment and attention now accorded to inducement prizes, there has been surprisingly little independent and robust assessment of the effectiveness and operations of such prizes. With impeccable timing, this book – *Technological Innovation and Prize Incentives* – by Luciano Kay, steps in to fill this gap.

In his book, Luciano Kay adds considerably to our understanding of innovation inducement prizes. He reviews the development and rediscovery of inducement prizes and then probes whether grand inducement prizes (also referred to as X prizes) actually result in additional or faster innovation and entrepreneurship. This is not a straightforward task: each prize has its own unique attributes and there are no readily available data sets to crunch. Luciano Kay carefully addresses the conceptual and methodological challenges posed in assessing prize incentives and painstakingly accumulates evidence and assesses counter-factual explanations. Two completed inducement X prizes are examined (the Ansari X Prize and the Northrop Grumman Lunar Lander Challenge), alongside field work in the United States and other countries to investigate the currently ongoing Google Lunar X Prize. The book highlights the diversity of motivations – beyond monetary rewards – that drive prize contestants and looks into how problem-solving activities are undertaken not only to win the prize but also to achieve other goals. We learn that many intermediate outputs are attained on the way to developing solutions to the prize challenge. Comparisons are made with conventional innovation practices and how competing for prizes relates to other innovation processes and objectives. Towards the end of the book, insights and improvement recommendations are offered to policymakers, sponsors and others seeking to offer or participate in future inducement prize competitions.

On a personal level, I am both pleased and honored to be invited to write the foreword to this book. I have witnessed how Luciano Kay has evolved this project from its initial formulation, research design and field work through to journal publications and professional reports, and now to this exciting and perceptive book. Luciano Kay reminds us of the importance of independent questioning and robust empirical research in understanding and evaluating the effects of innovation approaches and policies and in improving their design and implementation. By carefully analyzing the links between prize incentives and technological innovation, Luciano Kay opens up new avenues for dialogue about inducement prizes and also contributes significantly to the broader domain of technology and innovation studies, management, and policy assessment.

Philip Shapira
Professor of Innovation, Management and Policy, and Director,
Manchester Institute of Innovation Research, Manchester Business School,
University of Manchester, UK; and Professor, School of Public Policy,
Georgia Institute of Technology, Atlanta, USA

Acknowledgements

This book is the culmination of more than three years of arduous research work to both increase our understanding of the effect of technology prizes on innovation and, more generally, contribute original research to science and technology policy studies and draw lessons for science, technology and innovation policies. Many people and organizations have made this possible and I generally extend my gratitude to all of them. In particular, I am very grateful to the following.

Undoubtedly, this investigation of prizes would have been impossible without the valuable and unselfish contributions from key informants, interviewees, and leaders and members of the teams participating in the Google Lunar X Prize (GLXP), the main case study. In particular, I must thank William Pomerantz, X Prize Foundation's Director for Space prizes who helped in the initial steps of the process of data gathering, offered an interview to learn more about the GLXP and other competitions and invited me to attend the 4th annual GLXP Summit held in the Isle of Man (UK) in October 2010 – my participation in this event was very useful to collect additional data and meet some of the GLXP team leaders. I also thank Amanda Stiles from the X Prize Foundation, who provided access to the archives of the GLXP forum.

I also am very grateful to GLXP team leaders and members, particularly those who, despite their hard work and focus to compete and win the prize, were trustful and willing to collaborate with this research project and accepted interviews and visits to their workplaces. Other team members were also very kind and collaborated with interviews, filled questionnaires and contributed valued comments. All of them have been very generous to divert their attention from the competition and use their time to help me with this research. The author is especially thankful to William 'Red' Whittaker, Ruben Nunez, Andrew Barton, Neven Dološ, Robert Boehme, and Dumitru Popescu. Also to Rex Ridenoure, David Gump, Dillon Sances, Steve Murphy, Aad Eggers, Christian Bennat, Mario Kulczynski, Karsten Becker, Sebastian Rattay, and Simona Popescu. Other team leaders and members that kindly collaborated are Bob Richards, Richard Speck, Michael Joyce, Pete Bitar, Adil Jafry, Izmirov Yaminovich, Palle

Haastrup, Markus Bindhammer, Marc Zaballa, Alex Last, and Nikolay Dzis-Voynarovskiy.

A number of prize and industry experts also collaborated with interviews and key insights for this research, and I am very grateful to them: Ken Davidian (Director of Research at the FAA Office of Commercial Space Transportation – AST), Dennis Stone (Assistant Manager for NASA's Commercial Crew & Cargo Program), Jeff Greason (Founder and President of XCOR Aerospace), G. Thomas Marsh (retired Executive Vice President of Lockheed Martin), Gregg Maryniak (Vice President of Aerospace Science, St. Louis Science Center and X Prize Foundation's Advisor), and Norman Whitaker (DARPA's Deputy Director of The Transformational Convergence Technology Office).

I also must thank Prof. Philip Shapira, my mentor at Georgia Tech, for his constant encouragement and support to complete my research project on innovation prizes and write this book. Other senior colleagues kindly offered advice for this project too and I am thankful to them. I must mention Prof. Diana Hicks, Prof. Juan Rogers, Prof. Alan Wilhite, and Dr. Jan Youtie in this regard. Moreover, this project would not have been feasible without the appropriate economic support from the US National Science Foundation under Grant Number SBE-0965103 (any opinions, findings and conclusions or recommendations expressed in this work are those of the author and do not necessarily reflect the views of the National Science Foundation). I am also thankful to The IBM Center for the Business of Government for the support given for the preparation of the report *Managing Innovation Prizes in Government*, which seeks to translate research findings presented in this book into practicable recommendations for policy makers.

Finally and most importantly, I am very grateful to my family and friends who supported me through this project. They are always there, to give encouragement through the tough times and to celebrate all the accomplishments and great moments.

Abbreviations

AXP	Ansari X Prize
CMU	Carnegie Mellon University
COTS	commercial off-the-shelf
COTS	NASA's Commercial Orbital Transportation Services program
CRuSR	NASA's Commercial Reusable Suborbital Research program
DARPA	U.S. Defense Advanced Research Projects Agency
FAA	U.S. Federal Aviation Administration
GLXP	Google Lunar X Prize
HD	high definition
ILDD	NASA's Innovative Lunar Demonstrations Data program
IP	intellectual property
ISS	International Space Station
IT	information technology
ITAR	US International Traffic in Arms Regulations
JAXA	Japan Aerospace Exploration Agency
Mbps	millions of bits per second
MER	NASA's Mars Exploration Rovers
NACA	U.S. National Advisory Committee for Aeronautics
NASA	U.S. National Aeronautics and Space Administration
NGLLC	Northrop Grumman Lunar Lander Challenge
NGO	non-governmental organization
NSF	National Science Foundation
NTRS	NASA's Technical Reports Server
R&D	research and development
RASE	Royal Agricultural Society of England
SBIR	U.S. Small Business Innovations Research program
SME	small and medium enterprise
STEM	science, technology, engineering and mathematics
TLI	Trans Lunar Injection
TRL	technology readiness level
USAAF	U.S. Army Air Forces
VTOL	vertical take-off and landing
XPF	X Prize Foundation

1. Introduction

Prizes have long been used by public and private sponsors to elicit effort from individuals and organizations and attain diverse goals, including scientific discovery and technology development. Since at least the 18th century, for example, prizes have been used to encourage basic research by compensating research results with monetary rewards or medals (MacLeod, 1971; Crosland and Galvez, 1989; Brunt et al., 2008). This is the case of popular prizes such as the Nobel Prizes, which function as an incentive for scientists to achieve breakthroughs. There are also prizes offered ex-ante for the achievement of a certain technological target. These prizes typically offer a fixed, sometimes sizable monetary reward to the first entrant that achieves a prize challenge or to the entrant that progresses the farthest in a competition. This kind of prize may have been decisive to develop early innovations such as the marine chronometer and induce the initial development of the aviation industry in the 20th century (Davis and Davis, 2004; Maryniak, 2005; Mokyr, 2009). One of the most popular prizes of this kind has been the Orteig Prize for the first aviator to fly nonstop between New York and Paris (won in 1927 by Charles Lindbergh). During the last fifty years countless prizes have been offered in many fields attracting contenders and audiences with diverse interests (Best, 2008). But it has been a handful of successful global innovation prizes recently launched in the USA that revitalized the interest in this topic since the 1990s. These recent competitions include government prizes such as the $3.5 million DARPA Urban Challenge to develop autonomous robotic vehicles, and private sector prizes such as the $10 million Ansari X Prize for the first private reusable manned spacecraft and the $1 million Netflix Prize to improve Netflix Inc.'s movie recommendation system. Also older success stories such as that of the influential Orteig Prize were rediscovered in the 1990s and inspired the design of some of these modern prizes. The dynamism and broadly disseminated advances of those modern competitions suggested exciting opportunities in the use of prizes to exploit widely distributed knowledge and induce collaborative efforts to address critical issues, including technological innovation. In the USA this attracted the attention of policy makers and sparked further discussion between government stakeholders and some scholars (see, for example, NAE, 1999; NRC, 2007; Stine, 2009) which ultimately led to the enactment of new legislation to authorize federal agencies to use prizes broadly to accomplish their missions. Innovation prizes then made their way into the portfolio of policy instruments and their immediate implementation by USA government agencies such as the National Aeronautics and Space Administration (NASA) and the Defense Advanced Research Projects Agency (DARPA). More generally not only policy makers but also philanthropists, companies and the media have become increasingly interested in prizes due to their potential to induce path-breaking technological innovations or achieve related goals such as economic recovery, technology diffusion and the creation of innovation communities. Increasing numbers of proposals have also been put forward to implement prizes in fields as diverse as agriculture, medicine and nanotechnology to seek solutions to very challenging problems (see prize proposals by, for example, Horrobin, 1986; Kremer, 2000; Masters, 2003; Anastas and Zimmerman, 2007; Charlton, 2007).

To date, however, despite the long history of prizes as incentives for science and technology, their recent popularity and increasing policy interest, there has been little empirically based scientific knowledge on how to design, manage and evaluate innovation prizes. Prizes have generally appeared marginally in studies as variations of more traditional incentive mechanisms in the portfolio of policy options, but increasing interest by Science and Technology (S&T) policy scholars anticipates a better future for prize research. New prize-focused research programs investigate specific cases of historic and modern prizes empirically. This new empirical evidence will help to substantiate our knowledge on prizes which has generally been based on theoretical economic approaches (see, for example, Wright, 1983; de Laat, 1997; Shavell and van Ypersele, 1999; Newell and Wilson, 2005) and only to some extent on case studies that draw on anecdotal accounts or historical analysis (see, for example, Crosland and Galvez, 1989; Davis and Davis, 2004; Saar, 2006). Scholars have also participated in forums and workshops to inform the design of concrete prize initiatives and in communities of practice that focus on government prizes and disseminate case studies, prize announcements and related information (e.g. the USA Challenges Listserv).

Motivated by the increasing interest in prizes to attain diverse goals, this book seeks to close some significant knowledge gaps that call for further investigation on the design, implementation and evaluation of innovation prizes. The book presents the results of an empirical investigation of prizes and the means by which they induce innovation or other effects related to technological development. The focus is on four main aspects of prizes: the motivations of prize entrants, the organization of prize research and development (R&D) activities, the prize technologies and the overall effect of prizes on technological innovation. The investigation used a multiple case-study methodology and multiple types of data sources to investigate three cases of recent aerospace technology prizes: a main case study, the Google Lunar X Prize (GLXP) for robotic Moon exploration; and two pilot cases, the Ansari X Prize (AXP) for the first private reusable manned spacecraft and the Northrop Grumman Lunar Lander Challenge (NGLLC) for flights of reusable rocket-powered vehicles.

The book draws on prize literature insights and more general innovation literature, and addresses four main questions that are not only deemed relevant from the viewpoint of scientific inquiry but are also considered to have significant implications for policy making to design effective and more efficient prize competitions: (1) How do different types of incentives weigh in the overall motivation of different types of prize entrants? (2) What are the characteristics of prize R&D activities and how do they differ from traditional industry's R&D activities? (3) What are the characteristics of the prize technologies and how do they relate to the characteristics of prize entrants and their R&D activities? (4) Do prizes spur innovation over and above what would have occurred anyway? To be able to address these questions and probe corresponding propositions, the investigation introduces an innovation model that focuses on the prize competition as unit of analysis and articulates internal and external factors that can potentially explain the effect of prizes on innovation. To the author's knowledge, no framework or model of this kind has been offered by the academic literature to study the effect of prizes on innovation. This model is built upon six main dimensions identified in the prize literature, namely: prize design, motivation of prize entrants, R&D activities, technology outputs, characteristics of entrants, and the interplay between the prize and its context or technology sector. The model is used to pursue an iterative approach to empirical case study research. First the model is tested and improved with the retrospective study of two prize competitions of recent completion, the AXP and NGLLC. Second a refined version of the model is applied to investigate the main case study, the ongoing GLXP, and elaborate implications for theory, policies and future research.

The three case studies are among the most significant and popular modern prize experiences. The GLXP is a $30 million multi-year global competition organized by the X Prize Foundation (XPF) and sponsored by Google Inc. It was announced on September 2007 and has not found a winner yet. The GLXP requires participants to land a robot on the Moon, among other secondary goals, by December 2015. Thirty-five international teams entered the competition and participants from more than 40 countries have been involved. This prize has exceptional significance because it is an opportunity to gather valuable real-time data from ongoing R&D activities in a high-tech competition; it is interrelated with the strategic aerospace and defense industry sectors; and it has global reach, which offers the opportunity to observe the broadest impact of prizes. This is also an interesting case because the accomplishment of missions with goals similar to the GLXP may have other significant non-technological implications such as those related with geopolitical affairs (not specifically analyzed here). The AXP is a $10 million prize offered in 1996 for the first non-government organization to launch a reusable manned spacecraft into space twice within two weeks. It engaged 26 teams from seven countries. The USA firm Scaled Composites won this prize in 2004. The NGLLC, part of NASA's Centennial Challenges, is a $2 million multi-year prize offered for building and flying a rocket-powered vehicle that simulates the flight of a vehicle on the Moon. It involved 12 USA teams between 2006 and 2009. The USA firms Masten Space Systems and Armadillo Aerospace shared the prize money.

The analysis of these three case studies unveils the dynamics of these prizes and contributes a better understanding of the potential effects of prizes on innovation. Moreover, the book presents evidence of the complexity of the prize phenomenon and the uniqueness and specific features of each competition, revealing a big number of factors that influence the development of prizes and their ultimate outcomes. In the investigation of the ongoing GLXP, the book, however, does not seek to assess the performance of individual prize entrants and does not reveal sensitive information that might affect the strategies of competing teams. More generally, the investigation highlights the advantages and weaknesses of prizes under certain circumstances and provides insights for effective prize design and implementation. Many instructive methodological considerations also emerge throughout the analysis to inform further empirical prize research.

The book is organized as follows. Chapter 2 describes recent prize developments, reviews the more general literature that compares prizes with other incentive mechanisms, describes the types of prizes and discusses aspects related to the use of innovation prizes in government. Chapter 3 reviews the literature that is relevant to each of the four research questions and posits four corresponding hypothetical explanations. Chapter 4 discusses methodological aspects, introduces the innovation model to study prizes and describes the data and data gathering process. Chapter 5 presents the analysis and findings of the AXP and NGLLC case studies and presents considerations for model improvement and further research. Chapter 6 presents the analysis and findings of the GLXP case study. The findings are organized in subsections to address the six dimensions of the case study in three levels: the prize, the context and the prize entrants. Chapter 7 discusses case study findings and probes the anticipated effect of prizes. This chapter also seeks to advance the analysis and connect findings from the three case studies with the prize literature and insights of the broader innovation literature. Chapter 8 seeks to contribute new building blocks for the development of prize theory and presents policy and methodological considerations based on the findings of the three case studies. Chapter 9 makes some concluding remarks. The book also includes an Appendix with other useful data to understand the case studies.

2. Innovation, policy and prizes

2.1 PRIZE RENAISSANCE

In the last 20 years hundreds of prizes have been offered to accomplish very diverse goals. A report by McKinsey & Company estimates that all these prizes may be worth as much as $2 billion if the rewards offered by all types of prizes are added up. Inducement prizes (with a number of targets, including technology development) are those that have grown the most since 1991, offering rewards for about $236 million between 1991 and 2007 (McKinsey & Company, 2009). Scholars also note this growth in technology prizes and identify at least 38 innovation inducement prizes between 1990 and 2008, which includes at least 14 competitions organized by different USA federal agencies (KEI, 2008; Masters and Delbecq, 2008; Stine, 2009). Dozens of new private and USA federally funded prizes have been announced since then, but there is still no centralized information source to track this activity. This investigation had access to four recent studies that have sought to compile lists of the most significant prizes offered since at least the 16th century (Table 2.1). Three of these lists have broader coverage but prizes are not systematically categorized in all cases. There is also a compilation of government prizes limited to USA federally funded prizes. The top technology areas of prize implementation in terms of number of competitions vary depending on the data source. Aviation/ aerospace, climate/environment and medicine are among the top areas. Other more prominent areas include transportation (e.g. automotive), energy, defense, computing/software and chemistry. In spite of the variety of technology focus, a significant use of prizes in the aviation sector since the early 20th century and in aerospace since the 1990s suggests that prizes may be more effective in particular fields. The size of prize rewards varies considerably as well. The smallest technology development prize (excluding other forms of award) included in these reports was less than $100 offered for 'extracting sugar from native plants' by the Dutch Society for the Encouragement of Agriculture in the 18th century. The largest prize, with a cash purse of $53 million, was offered in 2004 by Bigelow Aerospace for 'transporting a five-person crew into orbit for 60 days, twice' (this prize expired in 2010 and was never claimed).

Table 2.1   Prize datasets recently compiled in selected literature

	Knowledge Ecology International (2008)	Masters and Delbecq (2008)	McKinsey & Company (2009)	Stine (2009)
Dataset content	204 awards and prizes	89 technology prizes	219 prizes worth $100 000 or more	14 US federally funded innovation inducement prizes
Coverage (years)	1567–2007	1567–2008	1769–2007	2004–09
Rewards range (cumulative total in parentheses)[a]	From $2.56 to proposal of $80 billion (>$80 billion)	From less than $50 000 to $53 million ($400 million)	From $100 000 to $30 million ($357 million)	From $250 000 to $10 million ($51 million)
Top technology areas (share of competitions in parentheses)	Medicine (18%) Aerospace (8%) Agriculture/ food (8%)	Aviation (20%) Medicine (11%) Transport (10%)	Climate/ Environment (11%) Medicine (9%) Aerospace (5%)[b]	Aerospace (43%) Energy (29%) Defense (14%)

Notes:
a.   Total estimate rewards comprise amounts for each edition of recurring prizes but do not include commitments to purchase inventions (values are estimates in US dollars for year of publication of dataset)
b.   Non-technology prizes (e.g. arts, literature, etc.) are excluded.

Source: KEI (2008); Masters and Delbecq (2008); McKinsey & Company (2009); Stine (2009).

The context in which modern prizes are implemented however is considerably different to the context of the early 18th century's contests. Most importantly recent successful prize competitions, including the X-prize series organized by the X Prize Foundation (XPF), are run by specialized organizations and have truly global participation enabled by the Internet and new communication as well as transportation means and virtual collaboration tools. The same technologies enable extensive media coverage and increasing visibility for the competitions, the participants and their sponsors as well. This has increasingly attracted both people interested in working to solve technical challenges and others willing to sponsor new prizes, such as philanthropists, government agencies and corporate officials that are interested in using prizes to meet diverse goals. In the USA many federally funded innovation prizes have been authorized since 2003. These prizes have been aimed at inducing research, development, testing, demonstration and deployment of technologies (Stine, 2009). Most of these prizes have offered cash rewards between $250000 and $10 million to solve challenges broadly related to the agencies' missions. NASA's Centennial Challenges program for example has used prizes to attract new entrepreneurs to develop aerospace technologies commercially. The US Department of Defense has used prizes to find innovative solutions in defense-related technologies. Among its most popular prizes are the Wearable Power Prize to develop long-endurance, lightweight power packs for war fighters, and the DARPA Grand Challenges to develop autonomous ground robotic vehicles. The US Department of Energy implemented the Bright Light Tomorrow Prize (L-Prize) competition to spur the development of ultra-efficient solid-state lighting products to replace the common light bulb. The US Department of Health and Human Services and other departments and agencies have implemented prizes as well.

Private companies also use prizes to improve their businesses or create prize-enabled enterprises. The companies InnoCentive[1] and NineSigma[2] for example have created online platforms where companies post challenges and communities of independent solvers work to find solutions and win cash rewards. Other examples are the $1 million Netflix Prize announced by Netflix Inc. (the film rental website that offers recommendations based on what customers watch) in 2006 to improve its movie recommendation system, and the $250000 Cisco I-Prize global innovation competition developed in 2010 to encourage collaboration among entrepreneurs in order to help identify new potential billion-dollar business ideas for Cisco Systems Inc.

Some private organizations have also been created specifically to administer prizes in the last 20 years. The XPF for example is an educational, non-profit corporation established in 1994 to inspire private, entrepreneurial advancements in space travel, and has sought to achieve such a mission by implementing prizes primarily with philanthropic support. This foundation organized the privately sponsored AXP and GLXP and NASA's NGLLC, all case studies in this investigation. Another non-governmental organization that has organized prizes is the CAFE Foundation, a research body that seeks to create and advance the understanding of personal aircraft technologies. This foundation has implemented, for example, the $1.65 million NASA-funded CAFE Green

Flight Challenge for the development of quiet, practical and green aircraft. CAFE Foundation has also organized some of NASA's Centennial Challenges prizes.

Reports and other scholarly works have attributed a variety of effects and technological impacts to prizes but none of them has engaged in the systematic, thorough investigation of the effects of prizes on innovation. This effect has generally been considered positive and may have included both technological development and commercialization, among others. The DARPA Grand Challenge 2005 for the development of autonomous vehicles, for instance, led to many technical accomplishments and remarkable improvement in several technologies related with autonomous driving (DARPA, 2006). NASA's Astronaut Glove Challenge 2007 for the development of spacesuit gloves induced technology commercialization, as the winner started a company and gained a contract to provide gloves to a spacesuits manufacturer (Stine, 2009). The AXP for the development of a suborbital spacecraft may have induced a total R&D investment by all prize participants of up to $100 million, which is ten times the cash purse offered by the sponsor (Newell and Wilson, 2005). The $1 million Netflix Prize announced by Netflix Inc. formed a problem-solving community of more than 34 000 developers worldwide (McKinsey & Company, 2009). The Cisco I-Prize engaged 2900 participants from 156 countries and received more than 800 new potential billion-dollar business ideas for Cisco Systems Inc. (Cisco, 2010).

2.2 TYPES AND STRUCTURE OF PRIZES

There are at least two types of prizes according to the achievement rewarded by the prize sponsors. *Targeted prizes* reward the achievement of challenges in the form of performance standards that must be met to claim the prize. These prizes have discrete success because there is (or there is not) achievement of the prescribed challenge and the characteristics of the achievement are more or less pre-specified by the prize rules. In the recent AXP, for example, the first participant to launch a reusable manned spacecraft into space twice within two weeks was declared winner of the prize. On the other hand *blue-sky prizes* (usually referred to as awards) are open-ended prizes that reward achievements that were not identified in advance. In this case achievement is a matter of opinion because the prize judges are allowed to know the winning achievement 'when they see it' (Scotchmer, 2005; Masters and Delbecq, 2008). The Nobel Prize is a well-known example of these awards.

This investigation focuses on targeted prizes that reward achievements associated with technological development (hereafter, innovation or technology prizes) and their effects on innovation. Technology prizes are generally organized as competitions in which participants are asked to attain a prize challenge. This challenge is defined in terms of a concrete technological problem to be solved, the deadline to find the solution or prize expiration date and, sometimes, the means to be used to solve the problem. The prize sponsor defines the prize challenge according to its interest in meeting certain goals and offers what is generally a sizable cash reward to the first participant to achieve that challenge. Modern innovation prizes generally have sponsors that contribute the cash purse, and organizers that manage the competition. They may be individuals, private organizations, government agencies or groups thereof. The sponsor and the organizer are the same entity in some cases. Unless otherwise indicated this investigation refers to both indistinctively. The prize challenge also establishes a relationship between the prize competition and certain technological fields and/or market segments and represents a technological gap that has to be reduced or closed by the participants. If no entrant achieves the prize challenge, the prize expires and the sponsor does not have to pay the reward. Prize entrants, or participants, are generally organized as teams of diverse composition and may include companies, universities, entrepreneurs or simply individuals that are attracted by the prize.

Innovation prizes can be structured and classified according to different criteria. This investigation refers to the classifications based on the following:

a. Required technological output:
    - Prizes for technology demonstration explicitly require building and demonstrating capabilities of a technology (e.g. NGLLC, AXP).
    - Prizes for technology-based achievements involve using unspecified methods to accomplish a feat or perform certain functions (e.g. GLXP).
b. Definition of the challenge:
    - First-to-achieve prizes define the challenge as a concrete technological goal that entrants have to achieve before the deadline to be eligible to claim the cash purse. The first entrant to achieve the challenge is considered the winner.
    - Best-in-class prizes define the challenge as a set of minimum standards of performance that entrants have to attain to be eligible to claim the cash purse. In this case the winner is the entrant that performs the best according to those standards. In best-in-class prizes there is typically a 'race' in which all participants come together

to compete for the cash purse. In this case the prize challenge may also be defined as a set of intermediate milestones or qualifying rounds to guide the effort of the participants and allow only the most qualified entries to be selected for a final round. If no participant achieves the minimum standards required by the sponsor in that final event, the prize is considered expired.

c. Number of awards:
   ● Winner-takes-all prizes award all the prize money to the winner of the competition.
   ● Multi-prize competitions offer cash rewards not only for the winner but also to the runners-up (e.g. second and third places).

2.3 PRIZES AND OTHER INCENTIVES FOR INNOVATION

Prizes are only one of many instruments used to stimulate technological innovation. Other much more widely utilized incentive mechanisms include the intellectual property (IP) system, research grants and R&D contracts. Most of our understanding of prizes is actually based on intuitive comparisons of prizes with those instruments and theoretical analyses aimed at finding what the optimal incentives are under specific circumstances. Little empirical evidence and some anecdotal accounts from recent prize experiences have also contributed insights to better understand the prize phenomenon.

Much emphasis of the literature has been on the debate about patents versus prize rewards because the latter may theoretically be able to solve one of the main defects of the patent system. The patent system grants inventors exclusive IP rights on their inventions and makes them monopolists because it excludes others from making, using or selling the invention. The problem with that monopoly is the deadweight loss that occurs when patent owners set prices for the invention that are above the marginal cost and produce less than the socially desirable quantity of output (Polanyi, 1944; Abramowicz, 2003; Scotchmer, 2005). In theory prizes can reduce or eliminate such deadweight loss by, rather than granting IP rights, awarding innovators with an amount of money equivalent to the social value of the innovation and requiring them to place their technologies in the public domain. This suggests that prizes may be more effective in areas where the social losses due to IP rights are likely to be higher (such as in the development of pharmaceuticals, computer software and recorded music and visual products) or where patents are expected to substantively distort cumulative innovation (Shavell and van Ypersele, 1999; Williams, 2010).

Both prizes and patents reward innovators for their research outputs but present four important differences in their practical application (Davis, 2002). First, prizes encourage the development of specific technologies that satisfy the requirements of the sponsor, and innovators bear the initial costs and risks of R&D. Conversely the IP system incentivizes innovation indirectly because innovators decide what to invest in according to their private information and assessments, and are punished by markets if they do not invest in the most valued technologies (Wright, 1983; Gallini and Scotchmer, 2001). Second, prize sponsors give innovators limited development lead times because they generally set a specific prize deadline to find a technical solution to the prize challenge. In the patent system companies control the development of their R&D activities and strategically advance or postpone deadlines according to their needs. Third, the prize reward is a fixed amount of money generally awarded to the winner of the competition and its value is linked to the goals and vision of the prize sponsor and not necessarily to the market value of the technology. The winning entry does not need to be the best available technical solution or the most affordable – it has to meet the prize sponsor's requirements. Conversely the value of patents is linked to the commercial merit of the technology and the ability of the inventor to introduce the new technology in the market. In other words the test of the new technology is performed in the market and not by a prize sponsor. Finally the role of prizes is limited to incentivizing the development of technologies indicated by the prize sponsors. The patent system is a more complex, decentralized decision-making system in which inventors decide the technologies they work on based on their private information and a coordination mechanism that signals the location of competences, eases technology trading and helps inter-firm collaborations (Penin, 2005).

There are also important differences between prizes and research grants. First, while prizes reward innovators for their research outputs, grants pay for research inputs. Prize sponsors pay only for research results in the technological field of their choice (i.e. the prize originates with a specific need) regardless of the cost of R&D activities of inventors or researchers. Conversely grant schemes operate as a self-selection system whereby researchers propose ideas to invest in and the funding agency decides whether to fund them. Second, prizes generally do not require the pursuit of any specific approach to achieve the prize challenge, which allows introducing novel approaches and unorthodox technological solutions. Research grant administrators are required to choose between different methods for achieving a particular goal, even when that includes the possibility of excluding nontraditional approaches (Kalil, 2006). Moreover, grant proposals generally describe the expected output of the research project whereas prizes do not anticipate what the ultimate characteristics of the winning entry are. Third, prize entrants are only paid upon the achievement of the challenge and thus bear the financial and R&D risks of their activities. Prize sponsors do not need in principle to monitor the activities of those that enter the competitions. Conversely grants provide researchers with up-front funding. Consequently the grant givers assume the R&D risks of the projects they fund and potentially face a moral hazard problem when the effort of grantees cannot be costlessly monitored.[3] Both grants and prizes are similar in the sense that non-monetary incentives operate in both types of mechanisms. The growth and professionalization of the scientific enterprise has led to a more complex reward system in which honorific rewards (such as peer recognition resulting in awards) have become increasingly important as well (Merton, 1973). Likewise, as discussed in the following paragraphs, prizes offer non-monetary incentives such as reputation and publicity for their participants.

Prizes also work significantly differently from R&D contracts. First, contracts establish ex-ante purchase conditions for the innovator that is chosen to procure R&D whereas prizes generally offer a fixed cash reward to the winner only. Second, while prizes make entrants bear R&D risks, procurement contracts typically stipulate terms that promote risk sharing between buyer and contractor. That risk sharing scheme depends upon the level of uncertainty that each project is likely to face, which gives origin to different types of contracts (Samuelson, 1986).[4] Third, contracts ideally require the buyer to be able to monitor R&D costs or innovation efforts (which is costly in practice) to prevent moral hazard problems and offer incentives for cost control by researchers (Wright, 1983; Rogerson, 1994). This gives advantage to other more decentralized incentive mechanisms such as prizes, in which the sponsor focuses on the design of the prize and the selection of the winner and not on the method ultimately used to achieve the challenge. Still one of the main applications of R&D contracts has been defense procurement. Similarly to prizes, these contracts are competitive when they include an initial phase of prototype competition and/or pre-selection.[5] Competition in R&D contracts however is sometimes limited to submissions of design proposals and, depending on the program, detailed studies or working prototypes (Rogerson, 1989). On the other hand contracts may not be efficient when it is difficult for the buyer to distinguish between high- and low-cost R&D performers on the basis of bids or costs submitted in the competitive phase. Prizes and contracts are similar in this regard as none of them guarantee that the superior idea, the most affordable product or the technology with the greatest commercial merit is actually chosen (Scotchmer, 2005).

Among the most notable works discussing the optimal application of these incentive mechanisms under different conditions are those by Wright (1983), de Laat (1997), Shavell and van Ypersele (1999), Scotchmer (1999) and Newell and Wilson (2005). Wright (1983) maintains that it is only exclusively private information of researchers that affects the optimal choice between incentives and that the potential advantages of prizes (compared to patents and research contracts) can be better appreciated in fields where the supply of research is inelastic and there are intermediate success probabilities of research projects. de Laat (1997) arrives at a similar conclusion yet maintains that information asymmetries about markets can only be used to justify patents rather than prizes when the R&D process is sufficiently competitive. Shavell and van Ypersele (1999) maintain that IP rights do not possess a fundamental social advantage over reward systems and that an optional reward system under which innovators choose between rewards and IP rights is superior to only IP rights. Scotchmer (1999) argues that it is optimal to grant patents in exchange for a fee (rather than using prizes) when sponsors do not have complete information on the benefits of the innovation because, by that means, the value of the reward is linked to its potential market value. Finally, assuming complete information about costs, benefits and probability of success, Newell and Wilson (2005) maintain that prizes, compared to other mechanisms, change the profits maximization function of firms and can induce different levels of research investment, offering an alternative option to policy makers to produce the optimal amount of research.

There have been other, sometimes neglected developments that can enhance the theoretical debate over the optimal choice of incentives under certain circumstances. While at the core of the comparison of prizes with patents is, for example, the choice of the inventor based on the availability of private and public information, empirical research has found that firms commonly rely on secrecy and not patents to protect their technologies (Levin et al., 1987; Cohen et al., 2000). In those cases factors other than the information on markets and research costs have to explain the decision of innovators about whether to file a patent, enter prize competitions or pursue alternative paths. Moreover, there are recent works that challenge the mainstream prizes versus patents discussion that assumes that firms use the patent system only in order to be granted short-term commercial monopoly rents. Penin (2005) for example maintains that in many industries most firms also use patents as strategic devices to trade technologies and to ease R&D collaborations, which suggests other, possibly more difficult to quantify considerations such as the value of partnerships or the acquisition of strategic technologies in the decisions about patents versus prizes.

Other scholars have also suggested, this time more intuitively, a number of factors that could make prizes more effective tools to promote innovation. Mowery et al. (2010) for example argue that prize competitions must specify precise output or performance targets to be effective and fair, and that the ability of entries in any competition to meet these targets must be readily verifiable, which may not be possible in some technologically diverse fields with many technical applications. Masters and Delbecq (2008) suggest that timing is key to the impact of prizes as technological progress changes achievable possibilities and socioeconomic conditions influence the desirability of those possibilities. Kalil (2006) suggests that prizes have to posit difficult yet achievable goals to be able to induce innovation. Newell and Wilson (2005) maintain that successful prizes have to offer a clear measure of success or target in a field where achievement is desirable but measurement has been lacking.

2.4 PRIZES AND S&T AND INNOVATION POLICY

During the last fifty years, countless prizes have been offered in many fields attracting contenders and audiences with diverse interests (Best, 2008). But it has been a handful of successful global innovation prizes recently launched in the USA that revitalized the interest in this topic since the 1990s. These recent competitions include government prizes such as the $3.5 million DARPA Urban Challenge to develop autonomous robotic vehicles, and private sector prizes such as the $10 million Ansari X Prize for the first private reusable manned spacecraft and the $1 million Netflix Prize to improve Netflix Inc.'s movie recommendation system. Also older success stories such as that of the influential Orteig Prize were rediscovered in the 1990s and inspired the design of some of these modern prizes.

The dynamism and broadly disseminated advances of those modern competitions suggested exciting opportunities in the use of prizes to exploit widely distributed knowledge and induce collaborative efforts to address critical issues, including technological innovation. In the USA this attracted the attention of policy makers and sparked further discussion between government stakeholders and some scholars (see, for example, NAE, 1999; NRC, 2007; Stine, 2009) which ultimately led to the enactment of new legislation to authorize federal agencies to use prizes broadly to accomplish their missions. In this manner innovation prizes made their way into the portfolio of policy instruments and their immediate implementation by USA government agencies such as the National Aeronautics and Space Administration (NASA) and the Defense Advanced Research Projects Agency (DARPA).

The America COMPETES Reauthorization Act passed in 2010 supports, among others, the use of prizes by federal agencies and departments. This legislation has provided all USA federal agencies with broad authority to conduct prize competitions and includes provisions for different aspects of prize design, implementation and oversight. It authorizes the use of prizes for one or more of the following: find solutions to well-defined problems; identify and promote broad ideas and practices and attract attention to them; promote participation to change the behavior of contestants or develop their skills; and stimulate innovations with the potential to advance agencies' missions. The legislation also allows agencies to enter into agreements with private, non-profit entities to administer a prize competition and requires reporting prize activity for each fiscal year. Another component of the USA actions to support the use of prizes is the Challenge.gov online platform with prizes offered by more than 20 departments and agencies. As of January 2011, there were over 55 announced competitions on that platform, with prize rewards that range from relatively small amounts of money ($200) to large amounts of money ($15 million) (Box 2.1 gives examples). Some communities of practice that focus on government prizes and disseminate case studies, prize announcements and related information have also been created (e.g. the USA Challenges Listserv).

To date government prizes still represent a small share of the government's efforts to promote R&D and innovation. For sake of illustration, a rough estimate shows that USA federally funded technology prizes represent only a negligible amount (about 0.05 per cent) of the total federal R&D spending and about 0.5 per cent of the federal tax preferences granted for R&D (Figure 2.1).[6] Compared to industry's R&D expenditures, the amount of rewards in government prizes is insignificant as well. Still these government prizes represent about 20 per cent of the total prize rewards offered since 2000.

The lack of empirical evidence and the short experience with government-sponsored prizes emphasizes the need for further research to inform decision-making processes on the use of prizes versus other incentives and the implementation of more efficient prize programs. Some scholarly research and recent prize experiences suggest that there are certain design and implementation features that make prizes a more appealing incentive mechanism to policy makers and prize advocates. First, the financial risk of the R&D activity in prizes rests with the competitors and their financiers as the monetary reward is only paid theoretically if there is a winner that achieves the proposed target. In conventional instruments such as R&D contracts and research grants, that financial risk rests largely with the taxpayers when the R&D activity is publicly funded. In practice there

BOX 2.1 RECENT PRIZE COMPETITIONS POSTED ON CHALLENGE.GOV (SELECTED EXAMPLES)

- *The Bright Tomorrow Lighting Prize ($15 million)* Sponsored by the US Department of Energy, the L Prize competition is aimed to substantially accelerate America's shift from inefficient, dated lighting products to innovative, high-performance products. The L Prize is the first government-sponsored technology competition designed to spur lighting manufacturers to develop high-quality, high-efficiency solid-state lighting products to replace the common light bulb.
- *The Progressive Automotive X Prize ($10 million)* The US Department of Energy, Progressive Insurance and the X Prize Foundation partnered to sponsor this prize, which was awarded in 2010. The goal of the prize was to inspire a new generation of viable, super-efficient vehicles that help break our addiction to oil and stem the effects of climate change. A project of the X Prize Foundation, the Progressive Automotive X Prize was an independent, technology neutral challenge for teams from around the world to compete in a multi-stage competition to produce clean, production-capable vehicles that exceed 100 miles-per-gallon energy equivalent (MPGe).
- *The Strong Tether Challenge ($2 million)* NASA sponsors this challenge in materials engineering as part of its Centennial Challenges. The tether developed by each team is subjected to a pull test and, in order to win the $2 million prize, the tether must exceed the strength of the best available commercial tether by 50 per cent with no increase in mass. A tether that can win this challenge would be a major step forward in materials technology. Such improved materials would have a wide range of applications in space and on Earth.
- *The Nano-Satellite Launch Challenge ($2 million)* Another NASA Centennial Challenges prize competition is to deliver two small satellites to Earth orbit in one week. Objectives of the competition include: (a) safe, low-cost, small payload delivery system for frequent access to Earth orbit;

(b) innovations in propulsion and other technologies as well as operations and management for broader applications in future launch systems; (c) a commercial capability for dedicated launches of small satellites at a cost comparable to secondary.

Note: Prize amounts in US dollars.

Source: Challenge.gov

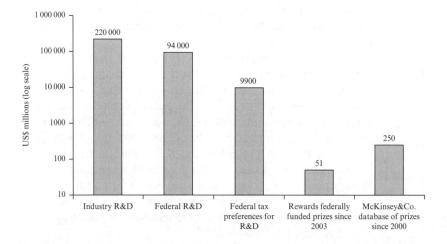

Notes: Industry and federal R&D spending as of 2006; federal tax preferences for R&D are an estimate of forgone revenues for 2006 (CBO, 2007); total rewards of federally funded prizes is an estimate shown as benchmark, for all prizes of this type between 2003 and 2009 (Stine, 2009), and do include competitions that offer procurement contracts as reward; McKinsey & Co. database of prizes is available in McKinsey & Company (2009).

Sources: NSF Science and Engineering Indicators 2008, otherwise indicated.

Figure 2.1 Amount of US federally funded prize rewards compared to US R&D spending, R&D incentives and total philanthropic prize rewards

are still some financial risks assumed by the sponsor (the costs of organizing the prize) and other potential risks if for example government property is used for prize technology testing and demonstration. Second, prizes can reduce the bureaucratic and accounting barriers to entry that accompany the grant and contracting processes and allow smaller companies to enter the R&D arena (Newell and Wilson, 2005). If we assume that smaller participants are likely to be less risk-averse than institutionalized competitors and pursue more technologically radical concepts (Nalebuff and Stiglitz, 1983) the reduction of bureaucratic barriers can lead not only to a more widespread participation but also to potentially more novel R&D approaches and technologies. Third, if they are properly implemented and result in fair and transparent competitions, prizes may prevent distortions in R&D spending caused by for example lobbyists in R&D contracts (Cohen and Noll, 1991). Fourth, prizes might be effective to target a full range of scientific and technological goals, including research, development, testing, demonstration and deployment (Stine, 2009) and other broader social and economic goals beyond technological development (NAE, 1999; NRC, 2007). Prizes might allow for example the participation of different social groups, including underrepresented groups, for technical training within a competitive environment. Competitions might also have a significant economic development impact by creating jobs and new businesses.

On the other hand there are a number of considerations that suggest a word of caution about the use of prizes in government. Most importantly, governments may lack information on the benefits and feasibility of inventions before they have been invented, which makes difficult the crafting of prize challenges and the calculation of monetary rewards (Kremer, 1998). This increases the risk of program failure because the more difficult to describe or measure objectively the innovation to identify the reward recipient, the more difficult to enforce the prize (Che and Gale, 2003; Newell and Wilson, 2005). A very instructive historical example is the Longitude Prize offered in 1714 by the British Parliament for the development of a method to measure longitude at sea. The inventor that built the most efficient technical solution was ultimately awarded the prize, but the scientific committee created to evaluate the innovation failed to judge it opportunely and objectively and thus the reward was paid late and only partially (Sobel, 1996). Moreover, this uncertainty on whether the prize will ultimately be awarded, for reasons such as bureaucracy, budget cuts or changes in administration, can weaken, if not eliminate, the incentives to compete (Macauley, 2005). This may also limit the scope of technological targets of government prizes, as prizes with longer lead times are likely to introduce more uncertainty. Also, while R&D contracts and research grants provide funding up-front to support early stages of technology development, prizes only reward the innovator upon the achievement of the prize challenge and hence create a barrier for small teams willing to participate. Finally government prize programs have to consider the most efficient use of funding. The administration of prizes can cost several times the amount of the actual cash purse. The total funding available for the DARPA Urban Challenge 2007 for example was $24 million: $12 million for the competition (including $3.5 million in prizes) and $12 million in seed funding to support a few qualified teams. Excessive budgets to fund experimental prize programs may lead to their termination, and policy makers might prefer traditional programs that are more expensive but involve more established or 'trusted' industry players as R&D performers. Potential solutions to this include alternative cost-bearing structures to alleviate the burden of costly programs. Government agencies may for example partner with other organizations to have their prize programs administered at no cost, as in the case of NASA's NGLLC. More recent USA legislation favors this kind of scheme and authorizes government agencies to accept external funds for cash prizes from other agencies or private organizations (Kay, 2011).

NOTES

1. http://www.innocentive.com
2. http://www.ninesigma.com
3. Still future grants are contingent on previous success and therefore grantees have a reason to be honest about their ideas and perform as proposed (Gallini and Scotchmer, 2001).
4. For example, cost-plus-incentive-fee contracts are more likely to be used when uncertainty is high, while fixed-price contracts are more likely to be used when uncertainty is low (Anton and Yao, 1990).
5. There have been for example U.S. Air Force defense contracts organized in three-step processes comprising proposal (including bid for initial production), pre-selection and prototype competition, and final selection and production (Rogerson, 1994).
6. The research and experimentation tax credit provides an incentive to undertake new research by giving firms a credit for expenses related to those new activities against the taxes they owe. In addition, R&D expenses that are not covered by the credit can be fully deducted from income as a business expense when incurred (CBO, 2007).

3. Key questions and hypotheses

3.1 INCENTIVES AND MOTIVATIONS OF PRIZE ENTRANTS

This investigation set out to examine the types of incentives offered by prizes and understand how different types of incentives weigh in the overall motivation of prize entrants. Findings in this regard are relevant for the design of successful prize competitions. To date most of our understanding of the motivations to participate in prizes has been based on the incentive provided by the monetary reward. In the scholarly literature the calculation of the appropriate prize reward has generally risen to the forefront but no accurate formula or algorithm has been provided to translate the theoretical concepts of private and social value of research and innovation into monetary amounts to determine prize rewards (Wright, 1983; Kremer, 1998; Shavell and van Ypersele, 1999; Abramowicz, 2003; Maurer and Scotchmer, 2004; Scotchmer, 2005; Wei, 2007). That is very important because theoretically, to induce R&D performers to pursue prizes rather than other options, rewards should reflect the social value of the inventions involved in the prize. Prize sponsors would prefer to pay only up to that amount if R&D costs were observable, but it is generally difficult for sponsors to observe and/or determine both the social value and the costs of R&D (not only for prizes but for other kinds of incentive mechanisms as well). Moreover, government and philanthropic sponsors are less likely to have information about the value of research and innovations than researchers and companies. Governments might have a better estimate of the social value only in certain cases that make prizes preferable to patents (for example when the social benefits of new medicines are known). Non-optimal prize rewards have various effects. If the reward is lower than the social surplus created by the invention, the incentive to invest in R&D would be inadequate and inventors would not be willing to compete for the prize. A very low reward may not even cover the costs of R&D. If the reward is equal to or exceeds the social surplus, the excessive incentive to invest in R&D would lead to inefficient duplication of investment.[1]

The literature has also discussed alternative rewarding schemes to address those issues related with information asymmetries in the use of prize incentives. To find optimal types of incentives, scholars have suggested for example schemes whereby the winner chooses between a monetary reward or a patent (Shavell and van Ypersele, 1999), a patent buy-out mechanism to harness private information on the market value of inventions (Kremer, 1998) and a prize reward conditional on a verifiable performance standard of the invention (Scotchmer, 2005). Some of these solutions had positive effects on innovation when implemented in the past. Kremer (1998) for example mentions the case of the Daguerreotype patent, purchased and placed in the public domain by the French government in 1839. This led to a worldwide adoption and innumerable technical improvements of such photography system. Other economics works have investigated what the optimal rewarding schemes are to make prizes efficient. In competitions with elimination stages for example, Rosen (1986) maintains that a distinguishable, large first-place prize is required in competitions with elimination stages to induce competitors to aspire to higher goals independent of past achievements. In multi-prize contests (i.e. with first- and second-place prizes), Moldovanu and Sela (2001) argue that the optimal-prize structure depends on the cost function of the prize entrants and that the right proportion between prize values depends on the number of entries, the distribution of abilities in the population and the curvature of the cost function of entrants as well. Lately a number of works (a few of them empirical) examined and brought attention to non-monetary incentives offered by prizes. For instance, in the empirical examination of the Royal Agricultural Society of England (RASE) prizes offered between 1839 and 1939, Brunt et al. (2008) found that a prestigious gold medal had greater entrant effect than cash rewards and that competitors viewed annual exhibitions of inventions as a powerful form of advertising. The same study found that cash rewards covered only around one-third of the total cost of the inventions exhibited by successful prize entrants, reinforcing the idea of the presence of other motivations beyond the cash purse. Other scholars contributed contrasting findings on these early prizes. In his investigation of the Royal Medals of the Royal Society of the 19th and early 20th centuries for example, MacLeod (1971) concludes that the medals induced competition yet, rather than encouraging fresh scientific discovery within British science as originally planned, they became a highly subjective means of personal recognition and legitimization of scientific paradigms. MacLeod also concluded that medals had to be combined with financial stipends if they were to be successful incentives for scientific discoveries. From their case studies of 20th century prizes, Davis and Davis (2004) concluded that reputation, credibility and visibility alone can provide the economic justification for a sponsor to design the contest and for contestants to enter. They also suggested that learning through technology spillovers and best practices diffusion is another potential motivation to participate in prizes. Davis and Davis (2004) and Saar (2006) also found that the potential market value of the prize technologies were a strong motivation for entrants in case studies of 19th and 20th century technology prizes.

Several other scholars also suggest the existence of those non-monetary incentives (Maurer and Scotchmer, 2004; Schroeder, 2004; Kalil, 2006; Anastas and Zimmerman, 2007; Culver et al., 2007). The kinds of incentives considered in the literature are diverse. Some works suggest for example that prizes can reduce bureaucratic and accounting barriers that accompany typical grant and contracting processes and thus attract smaller firms or independent researchers. Most importantly, some argue, the openness of the prize process may allow the participation of unconventional innovators, that is, individuals and organizations that are not generally involved with the development of the prize technologies and use non-traditional approaches to pursue the prize challenge (see, for example, Schroeder, 2004; Newell and Wilson, 2005; Culver et al., 2007). Direct observation of recent prize competitions also hinted at the presence of non-monetary incentives. Program managers have identified cases in which for example the costs of development in prize entrants' projects exceed the amount of the cash purse considerably (see, for example, Davidian, 2007), calling into question the importance of the monetary reward and suggesting the presence of other types of incentives.

Certainly there may be cases in which prize entrants are not aware of the theoretical private/social benefit concepts addressed by economists in the study of incentive mechanisms. Scientists or engineers that participate in prizes for example are unlikely to make their decisions about participation based on those terms. Instead their decisions are possibly based on other personal or professional factors such as career growth, personal finances and other personal accomplishments. Even the exclusive consideration of monetary incentives may be misguiding in the interpretation of the primary motivations of entrants. The case of the Kremer Prize of 1977 is illustrative in this regard. Paul MacCready, aeronautical engineer, decided to compete in (and ultimately won) the prize because he owed money for an amount equivalent to the cash purse. He explained in a later account:

> I did recall, with no special emphasis, this £50000 prize that Henry Kremer had put up 17 years earlier. And then, one day I happened to notice that at that time the pound was worth just two dollars. Suddenly, this great light bulb just glowed over my head: the prize was $100000, my debt was $100000. There just may be some interesting connection between these two. So my interest in

human-powered flight zoomed up to high level, and I fussed away at it, and eventually it worked. (Academy of Achievement, 1991)

The lack of a better understanding on the incentive effect of prizes raises the first key question (Q1): How do different types of incentives weigh in the overall motivation of different types of prize entrants? To be able to address this question this investigation introduced two assumptions on the classification of incentives and entrants. The first assumption is that there are at least two types of incentives: (a) prize incentives, defined as those offered exclusively by the competition and which would not exist if the prize was not announced; and (b) technology incentives, defined as those that are linked to the value of the prize technologies. By definition the former can be set by the prize sponsor with certain precision to produce the desired effects. These incentives include for example monetary incentives (e.g. cash prizes or bonuses) or non-monetary incentives (e.g. prestige, publicity or reputation for prize entrants). Technology incentives on the other hand are linked to the market value of the prize technologies and/ or the benefits that result from the introduction of the prize technologies for own use. These include for example potential revenues from commercialization or cost savings obtained by the exploitation of the technologies in own performance improvement. To better understand the incentive effect of prizes, this investigation also considered that the individuals that enter competitions may be actually motivated by other reasons not directly related with the prize or the market value of the technology. In particular this analysis considered intrinsic motivations, i.e. participation driven by an interest or enjoyment of the task itself (Ryan and Deci, 2000) or altruistic motivations, related with social values and the desire to help others.

There is also a second assumption that facilitates operationalization and helps to probe the ability of prizes to attract outsiders to technology development. This investigation considered that there are *unconventional* entrants that are not generally involved or familiar with the development of the prize technologies, and *conventional* entrants that comprise individuals and organizations generally involved with the development of prize technologies. The attribute that defines such involvement in this investigation is the industry experience that individuals and organizations have with the prize technologies. In the case studies examined by this investigation, groups of individuals with significant work experience in space agencies/industry have been considered conventional entrants. The rest of the prize entrants are considered unconventional.[2]

Prior research and intuition suggested at least two main potential explanations for the relationship between type of incentive and type of entrant. This investigation considered that prize entrants might factor out only the value of the cash purse offered by the prize in their decisions to participate. This alternative, which resembles a simplified economics perspective to incentives, assumes that would-be entrants evaluate the prize money and the risks of prize participation, compare them with alternative strategies based on their private information, and only enter the competition if they foresee a profit. In the examination of how different incentives weigh in the decisions to compete, this alternative represents a null hypothesis as only one type of incentive is perceived and no variation would be measured. There is a more elaborate explanation that includes other literature insights. Entrants might have unique characteristics that affect their perception of both monetary and non-monetary incentives and their decisions to participate. For instance, individuals not generally involved in technology development may perceive in prizes the opportunity to participate and learn, an opportunity that they would not otherwise have access to. Other individuals and organizations already involved in technology development may perceive in prizes the opportunity to create synergies with their ongoing activities and use their expertise to win the prize and/or develop valuable technologies for their own projects. Hypothesis 1 (H1) adopts the perspective of this second explanation and posits that there are entrant-level factors and non-monetary incentives that influence prize participation. This hypothesis also maintains that the individuals and organizations that enter prizes have different perceptions of the incentives offered by prizes. The alleged ability of prizes to attract outsiders suggests the experience with the prize technologies as a distinctive attribute in entrants. Therefore H1 anticipates that, for any given technological field and its general context, more significant prize, monetary and non-monetary incentives are more likely to induce the participation of unconventional entrants and more significant technology incentives are more likely to induce the participation of conventional entrants.

3.2   PRIZE R&D ACTIVITIES

In technology prizes entrants have to perform some kind of R&D activities to achieve the challenge before the prize expiration date. Our understanding of the organization of those activities is mostly intuitive and limited to very general (though still instructive) features of R&D observed in recent prize experiences. One of the most intriguing questions that remain to be addressed is whether the intense, competitive environment of prizes induces R&D activities that differ from traditional industry practices. Moreover, strict deadlines to find a solution to the prize challenge and the lack of up-front funding that entrants typically face in prizes also suggest potential differences with other instances of technology development. Still we shall consider that prizes are not only instances of competition but also cases of collaboration and communication. Awards in general (and not only technology prizes) may serve the function of what economists call 'communication' as they bring disparate players into informed contact with one another so that mutually beneficial transactions occur among them (English, 2005). Simultaneous competition and collaborations between problem solvers have been also found to be key for individual innovators to succeed in recently launched virtual online prize platforms (Bullinger et al., 2010; Hutter et al., 2011). This phenomenon of competition and formation of research communities to advance specific technologies and solve concrete technical problems has been also observed in recent competitions such as the DARPA Challenges and the Netflix Prize (DARPA, 2006, 2008; McKinsey & Company, 2009). Low entry barriers to prizes given by registration requirements that are easy to meet may have some desirable effects on prize participation and R&D activity. In addition to unconventional entrants, prizes might also benefit from the participation of small players that are less risk-averse and more likely to pursue technologically radical concepts than institutionalized competitors which generally compete for grants and contracts (Nalebuff and Stiglitz, 1983). Likewise broader prize participation might encourage unconventional partnerships between entrants and other entities and contribute new ways to organize R&D (Culver et al., 2007).

Much more has been discussed about the ability of prizes to induce R&D effort and leverage investment. Theoretically larger prize rewards may lead to more vigorous R&D races and shorter achievement times for technology development considering entrants that start from similar position (Grishagin et al., 2001). Some argue that ex-ante fixed rewards, deadlines and technology specifications have the potential to induce very focused R&D efforts (Newell and Wilson, 2005). The drawback to these potentially increasing activities is the potential duplication of R&D when prizes engage large numbers of participants, because there tends to be only a finite number of innovative ideas for any given technology problem at a given time (Maurer and Scotchmer, 2004; Newell and Wilson, 2005). In the same vein some maintain that free entry to competitions is not optimal and therefore entrants have to be taxed with an entry fee. Otherwise the individual R&D effort of prize entrants would weaken when there is an increasing number of competitors and entrants perceive fewer chances to win the prize (Taylor, 1995; Fullerton and McAfee, 1999). There is also some empirical evidence on the ability of prizes to leverage funding. Brunt et al. (2008) found for example that monetary rewards only offset one-third of the costs of technology development in RASE prizes. Schroeder

(2004) also estimates that the AXP induced an investment from entrants 40 times the size of the cash purse, and that the DARPA Challenges induced investments up to 50 times the size of the cash purse. Recent prize experiences such as the NGLLC may have also shown that prize entrants spend several times the cash purse to achieve the prize challenge (see for example Davidian, 2007).

The interest in the organization of R&D activities in prizes has been particularly raised by the exceptional performance of entrants in recent prize competitions. These entrants have had significant technological achievements with low budgets and sometimes no previous experience in the field. Let us consider for example Armadillo Aerospace from Mesquite, Texas. This is a team created in 2000 by a small group of mostly Information Technology (IT) professionals to enter the AXP and develop a spacecraft with suborbital flight capabilities. The team also entered the NGLLC in 2006 to develop vertical take-off and landing rocket vehicles. The team spent at least $3.5 million in its R&D program (a relatively small amount for this kind of space program) and won two prizes for $850 000 (Armadillo Aerospace, 2008). In those competitions the team introduced sophisticated computer controls for its vehicles and contributed to establish new standards of reusability, operation, speed of development and efficiency (NASA, 2009a).

To the author's knowledge the unique characteristics of those prize R&D activities and the extent to which they differ from traditional industry practices have not been investigated. There are at least three basic characteristics of prizes that can be examined to gain insights into this topic. First and foremost prizes generally do not provide up-front funding to perform R&D. That might create conditions that differ from R&D undertaken in instances of procurement contracts, research grants or even corporate new product development, which generally do have some (or all) funding up front to perform R&D. This should be particularly relevant for projects that involve technologies of expensive development and that require access to expensive facilities or equipment for their development. Recent prize experiences have sought to overcome potential issues in this regard with the implementation of hybrid approaches that include monetary rewards and seed funding to support entrants' R&D activities (see for example DARPA, 2008). Still, prize entrants are likely to focus their effort on a single and discrete goal rather than having to pursue a continuous activity with multiple projects or customers and maintain a considerable industry-like infrastructure, which ultimately might allow implementing low overhead R&D organizations. In the second place prizes also pose strict deadlines to come up with a technological solution to the prize challenge. Deadlines and shorter lead times are not an exclusive feature of prizes since time is generally considered a key resource for R&D teams working in all competitive environments (Waller et al., 2001). Deadlines however may be interpreted in two different ways: as the available time to achieve a specific goal, or as part of the overall goal that team members must work toward achieving (Locke and Latham, 1990; Karau and Kelly, 1992). By definition the deadline in prizes is part of the prize challenge and thus is defined by the sponsor at once for all entrants. In other words, regardless of the organization of their activities, entrants have to produce the technical solution within the given time frame to be able to claim the prize. In other contexts, such as commercial new product development or government procurement contracts, the R&D performer sets and/or negotiates deadlines and may even be able to postpone deadlines based on its own strategy and/or available resources. Last but not least the often cited ability of prizes to induce R&D effort from unconventional entrants is also intriguing. These entrants comprise individuals and organizations that are generally not involved with the prize technologies and might bring to the competition fresh ideas, unique perspectives to the prize challenge and unorthodox R&D approaches.

This investigation builds upon those basic features of prizes and literature insights to address a second key question (Q2): What are the characteristics of prize R&D activities and how do they differ from traditional industry's R&D activities? Intuitively this investigation set out to probe the effect of two of the unique features of prizes on the organization of R&D activities in this competitive context: the lack of up-front funding and the strict prize deadline. These are not only unique features of prizes but also conditions set by prize sponsors when they design competitions. The prize expiration date is generally explicitly stated in the prize rules. They define the maximum development lead time available to produce solutions to the prize challenge. The time ultimately available to prize entrants however varies significantly depending on their entry point in the prize timeline, a phenomenon that is further discussed in following chapters. The significance of the lack of up-front funding on the other hand varies according to how expensive is the achievement of the prize challenge. Prize challenges that involve potentially significant R&D efforts to produce a technological solution create a wider funding gap for entrants. The importance of this gap however also depends to a great extent on the resources available to entrants.

Short development lead times and the lack of up-front funding might become significant constraints in the competitive context of prize competitions. The need for faster achievement times and the lack of funding might push entrants to, for example, introduce simpler technologies to be able to come up with solutions faster and at a lower cost. Furthermore, entrants might seek to shorten lead times and reduce costs by drawing upon/combining existing technologies rather than developing new technologies to achieve the prize challenge. If entrants cannot respond to time and funding constraints with alternative design criteria or using existing technologies, they may engage in increasing collaborative efforts that help in accomplishing the prize challenge. Based on those insights this investigation focused on three potentially unique characteristics of prize R&D activities: idea sources and designs introduced by entrants,[3] the extent to which entrants draw upon existing technologies to achieve the prize challenge, and the extent of R&D collaborations. This focus is fairly generic to allow comparisons with other instances of R&D activity, but also reasonably specific to allow operationalization and hypothesis probing. It is also plausible that time and funding conditions do not represent a constraint for entrants and therefore these factors do not cause significant differences between prize R&D and activities performed in other contexts such as corporate and government R&D.

Hypothesis 2 (H2) posits that there is that kind of relationship between time and funding conditions in prizes and the characteristics of the induced R&D activities. It anticipates that for any given technological field, shorter lead times and more significant funding requirements posed by the prize lead to simpler technological designs, more significant reliance upon existing or standard technologies and more collaborative R&D efforts (H2). In this context the definition of simplicity is associated with the number of parts and their interconnectedness in a technological system; that is, simpler designs have fewer and less interconnected parts than complex designs. Existing technologies are those considered readily available, commercially or by other means. More collaborative efforts are those that involve an increasing number of actors (individuals and/or organizations) and linkages between them and the prize entrants.

3.3   PRIZE TECHNOLOGY OUTPUTS

Not much has been written on the characteristics of the technologies produced by prize entrants to come up with a solution to the prize challenge. This investigation refers to technology outputs in a broad sense to include outputs that entrants start producing as soon as they enter the competition or complete before prize expiration if they have already ongoing projects. These outputs include diverse forms such as conceptual designs, models, prototypes and actual products or services. Prize entrants may use their technologies in their own projects or introduce them into new or existing markets. Although the winning entry is generally the most visible output of prizes, runners-up and other participants may also contribute significant technological developments.

Under some circumstances prizes may not be efficient if they lead to technological solutions that are not valuable from the point of view of the prize sponsor. One of the most cited advantages of prizes is their ability to offer greater scope for unexpected solutions and for solutions arising from unexpected sources compared with conventional innovation policy instruments (Kalil, 2006) which is the result of, some argue, the participation of unconventional prize entrants that are more likely to introduce novel technologies and methods. But even in those cases prizes may not be so efficient as to induce the development of technologies of certain quality or performance. In other words prizes cannot guarantee that entrants with the best ideas are motivated to participate or, if that happens, that the best idea is selected or implemented at the minimum cost (Scotchmer, 2005). On the other hand prize sponsors might use more strict technical specifications for the prize challenge to specifically induce the development of technologies they consider more valuable, but that implies not taking full advantage of the alleged ability of prizes to tap into the creativity of participants to produce innovative solutions. Moreover, in some cases more strict challenge definitions might not be applicable. If it is difficult to anticipate what combination of ideas or resources will be required to address a particular problem, prize sponsors might come up with inadequate challenge definitions and therefore potentially ineffective prizes (Gans and Stern, 2010).

Prizes may even fail and not produce any valuable technological output at all. A potential pitfall is precisely related with strict technical specifications of the prize challenge. Too specific challenge definitions may prevent some significant innovations from being judged winning entries and being rewarded because they are not up to the prize's exact requirements (Kremer, 1998; Masters and Delbecq, 2008). Moreover, the prize challenge might be too ambitious for current-day capabilities and/or too expensive to achieve under current economic conditions, in which case the prize would fail to find a winner (Macauley, 2005). On the other hand, given the sequential and cumulative nature of innovation, prizes might be able to induce technological advancements only up to a certain point and fail to induce subsequent superior inventions if the proper incentives are not offered (Gallini and Scotchmer, 2001). This also relates with the ability of prizes to induce commercialization and widespread adoption of technologies. Prizes may provide adequate incentives to make an invention occur, yet the invention may still never be applied or reach the market for commercialization if prize entrants only target the prize challenge and lack sufficient incentives to develop the invention commercially (Kieff,

2001; Wei, 2007). Moreover, the prize technologies may lack commercial merit if they are very unorthodox and represent a radical departure from traditional technology and business models.

Beyond all these considerations there is still little empirical evidence on the kind of technologies produced in prizes and their determinant factors. An early examination of prizes awarded by RASE in 1851 called into question their ability to induce valuable innovations as only two out of 170 awarded medals rewarded extraordinary novelty or utility (Sidney, 1862). Conversely a more recent study of RASE prizes awarded between 1839 and 1939 found that prizes were correlated with patenting activity in the field and had a large effect on the quality of the invention as measured through patent renewal fees (Brunt et al., 2008). Recent case studies of prizes implemented during the 20th century found that competitions motivated inventors and stimulated innovations to achieve increasing performance goals in fields related with the prize technologies. Even when there was no immediate commercial gain, prizes were valuable to achieve the most efficient innovation outcome, improving designs that were not entirely new (Davis and Davis, 2004). Prize sponsors have also attributed positive effects of prizes on technology development. The DARPA Grand Challenge 2005 for autonomous vehicles for example led to many technical accomplishments and remarkable improvement in several technologies related to the prize challenge (DARPA, 2006). The NGLLC has also induced the development of sophisticated instruments and the achievement of new standards of reusability, operation, speed of development and efficiency (NASA, 2009a).

The third key question addressed by this investigation is about the characteristics of the prize technologies (Q3): What are the characteristics of the prize technology outputs and how do they relate to the characteristics of prize entrants, their R&D activities and parameters of prize design? Since the prize technologies can range from ideas and concepts to successfully implemented devices or technological systems, to address such a question this investigation focuses on the degree of novelty and maturity of the technologies advanced by entrants. These two attributes can provide significant insights on the innovativeness and potential use or implementation of the prize technologies. Innovation is generally defined as the creation and implementation of new or improved products, processes or organizations (OECD/Eurostat, 1997). This investigation considers the degree of novelty to be relative to industry's current-day products, processes and organizations, and it is measured in a scale that comprises current-day, derivative/improved and breakthrough technologies. Using standard definitions, derivative/improved technologies are those more efficient or affordable versions of current-day technologies,

Table 3.1   Technology Readiness Levels (TRL) applied to space engineering

TRL	Definition	Phases	Activities to advance the technology
9	Actual system *flight proven* through successful mission operations (ground or space)	Production and deployment	Implementation
8	Actual system completed and *flight qualified* through test and demonstration (ground or space)		
7	System prototype demonstration in a space environment (ground or space)	Systems development and demonstration	Development
6	System/subsystem model or prototype demonstration in a relevant environment (ground or space)		
5	Component and/or breadboard validation in relevant environment	Technology development	
4	Component and/or breadboard validation in laboratory environment		
3	Analytical and experimental critical function and/or characteristic proof-of-concept	Research and feasibility demonstration	Research
2	Technology concept and/or application formulated		
1	Basic principles observed and reported		

Source: Author based on Mankins (1995).

while technological breakthroughs are those able to generate a paradigm shift in the science and technology and/or market structure of the industry sector (Garcia and Calantone, 2002). Implementation refers to the deployment or commercialization of technologies (or services based on them) and use of the technology for own performance improvement (in this case for example in the attempts to win a prize). In engineering projects the concepts of novelty and technology implementation can be also linked to Technology Readiness Levels (TRL) (Table 3.1). TRL levels are a systematic metric for technology maturity that allows the consistent comparison between different types of technologies (Mankins, 1995). Novel ideas and concepts are linked to research activities (i.e. pure research) at low TRL levels; proof-of-concepts, tests and capability demonstrations are linked to medium TRL levels; and the advancement of technologies for their actual deployment or commercialization are linked to higher TRL levels. Government agencies and major private companies use this measure to assess the maturity of evolving technologies such as materials, components and devices prior to incorporating them into a system or subsystem. In space projects, for example, TRL levels indicate the maturity of the technology in relation to being acceptable for launch, or mission-ready. TRL levels also relate to the risk and/or uncertainty that the use of the technology involves: the lower the TRL, the greater the risk/uncertainty. TRL levels do not correlate to any specific project management principle or guideline (Sauser et al., 2005). This investigation uses TRL levels to assess the prize technologies and their relationship with current-day technologies.

In the consideration of the factors that may affect the degree of novelty and the actual implementation of the prize technologies, there is the appealing notion by which unconventional or new-to-industry entrants are more likely to introduce more novel ideas and approaches as they draw on perspectives that are atypical to the technological field the prize focuses on. The caveat might be that these entrants may effectively contribute novel ideas but, since they are not familiar with the prize technologies, they would also lack the skills or resources to further develop those ideas and come up with a solution to the prize challenge. Conversely conventional entrants would have access to knowledge, skills and professional networks that might enable implementation and commercialization of technologies. These entrants may even have pre-existing projects related with the prize technologies and seek to develop these technologies for those other projects. The technologies developed in prizes are also likely to depend on prize design features such as the definition of the prize challenge. Attributes such as technical specifications and the time allowed to find solutions to the challenge may enable (or not) the introduction of more novel ideas and longer-term development processes, respectively. Finally the market incentives associated with the prize challenge might influence the technologies that entrants work on as they represent additional incentives to the reward. Hypothesis 3 (H3) is then introduced to uncover the factors that determine the prize technology outputs, probe the literature assumption by which unconventional entrants are more likely to come up with innovative solutions, and better understand whether prizes can be designed to, for example, target certain individuals or communities of problem solvers and incentivize the development of different kinds of technologies. It posits that, for any given prize challenge definition, unconventional entrants are more likely to introduce novel technologies (H3a), and the more significant the technology incentives associated with the prize challenge, the closer to implementation/commercialization are likely to be the prize technologies (H3b).

3.4   THE OVERALL EFFECT OF PRIZES ON INNOVATION

The overall effect of prizes on innovation is after all the most important aspect addressed by this investigation. Much has been said about the virtues of prizes to produce multiple effects including technological innovation, but no significant empirical evidence has been contributed by scholarly research to be able to explain the real effects of this kind of incentive mechanism. Moreover, the literature is not precise on the nature of those effects and what their causal factors are.

In one of the earliest references of the academic literature to the potential of prizes to induce innovation, Michael Polanyi suggested that prizes might increase the amount of industrial research that is published because industrial laboratories would become eager to claim potential rewards. This would eventually contribute, at least indirectly, to some important technical innovations (Polanyi, 1944). More recently scholars and experts started to elaborate on the potential of prizes to accelerate innovation or speed up development in certain technological fields (see, for example, Anastas and Zimmerman, 2007; Culver et al., 2007; Masters and Delbecq, 2008; McKinsey & Company, 2009). Today's mainstream discussion refers to a wide range of abilities of prizes to, for example, mobilize human capital, attract investment, induce focused problem-solving activities, raise industry and public awareness and more generally spur innovation. Others have suggested that prizes can also change the direction of the innovation pathway or 'focus innovative efforts on problems for which solutions otherwise do not seem to be forthcoming' (Davis and Davis, 2004). Scholars also maintain that prizes may also be able to motivate the last effort to come up with a technical solution that is already under development (Saar, 2006), be efficient instruments to induce technological breakthroughs (Mowery et al., 2010), and serve as an innovation catalyst by lowering entry barriers to markets, thus enabling the participation of a much more diverse range of players (Culver et al., 2007). Some have observed that the mere announcement of prizes in the past has induced immediate considerable activity to find technical solutions to the prize challenge (see, for example, Sobel, 1996; Kessner, 2010).

The evidence of such effects on innovation can be revealed in different forms. Scholars maintain that the potential technology accomplishments induced by prizes may include new inventions, new applications, performance improvements and technology diffusion (NAE, 1999). Moreover, those accomplishments can include both incremental and radical innovations. Recently compiled prize datasets account for both improvements and technology breakthroughs in diverse technological fields, ranging from the creation of methods to preserve food in early prizes to the recent AXP's achievement linked to private commercial spaceflight (see, for example, Masters and Delbecq, 2008; McKinsey & Company, 2009). Also based on its experience with prize design and implementation, the XPF has suggested that short-term prize competitions with relatively small rewards can be aimed at inducing incremental technological changes, and that medium- to long-term competitions with more significant cash purses can be aimed at inducing revolutionary changes and breakthroughs (Pomerantz, 2006).

But all this discussion on the ability of prizes to spur technological innovation has developed much faster than the actual empirical evidence on prize effects. Recent empirical studies include among others the work by Brunt et al. (2008) on the RASE prizes for the development of tools and implements for agriculture between 1839 and 1939. These scholars conclude that those early prizes influenced positively the direction of technological effort and its quality, as indicated by the increasing patenting activity for the prize technologies. Interestingly a very early account on the RASE prizes by Sidney (1862) points out that those prizes actually did not have more effect than the traditional exhibitions and industry competitions and generally speaking prizes only misdirected the effort of people. The case studies by Davis and Davis (2004) show that prizes had an important role in the development of motorized flight, human-powered flight and energy efficient refrigerators during the 20th century. Although in some cases prizes did not produce technologies with immediate commercial merit, they stimulated innovation and further development in related technologies and industries. Based on his investigation of five case studies of 18th and 19th century prizes, Saar (2006) also maintains that non-traditional, unexpected technical solutions were a common feature of those prizes with the caveat that in some cases the winning entries were possibly already under development when the prizes were launched.

If one has to summarize the growing interest in prizes into a single question, the focus would be most likely on whether all those potential and observed effects can actually be attributed to prizes. In other words (Q4): Do prizes spur innovation over and above what would have occurred anyway? This is possibly the most relevant aspect of the study of prizes for at least two reasons. First, there has been much discussion about the virtues of prizes to advance technologies and spur innovation, but there is almost no empirical evidence in this regard. Second, there is a growing interest to make a more widespread use of this type of instruments to promote innovation and achieve other related goals, and further understanding of prizes and their effects based on empirical evidence and systematic examination can inform the decisions to design and implement prizes. At a glance, addressing that kind of counterfactual question might become troublesome but, while the use of counterfactuals has been generally criticized due to their alleged ambiguity, determinism and problematic implementation, counterfactual conditionals also involve causal implications and are always implicit or explicit components of the analysis of causality (Roese and Olson, 1995; Broda-Bahm, 1996). This investigation ultimately shows that in practice addressing a counterfactual question can yield a more enlightening debate and be effective to guide research inquiry aimed at uncovering the real potential of prizes to induce innovation. Undoubtedly prizes do induce some type of R&D activity and might help in advancing technologies. The question is whether the same outcomes would be observed if the prize was not announced and only more traditional incentive mechanisms were used for technological development. Therefore to understand the kind of prize-specific effect on innovation in a given field, we shall consider not only current-day technological capabilities and the nature of the challenge posed by prizes but also other industry developments that relate to the prize target. Evidence of innovation would be for example (a) technology advancements toward the prize target in a prize that requires increasing technological capabilities over current-day technologies (which concerns the ability of prizes to accelerate technological development) and/or (b) breakthroughs or unexpected prize technology outputs from the point of view of technological capabilities anticipated by industry forecasts and roadmaps (which concerns the ability of prizes to induce the development of technologies that otherwise do not seem to be forthcoming). Other related evidence might also be revealed by the influence of prize technologies on industry developments.

This investigation anticipated that prizes do induce innovation over and above what would have occurred anyway and focused particularly on innovation effects as they relate to prize design parameters that can be set by sponsors. Hypothesis 4 posits that for any fixed technological field and its general context, more significant prize incentives, more significant technology gaps involved in the prize challenge and more open-ended challenge definitions are more likely to induce innovation (H4). This hypothesis involves two key prize design parameters: the prize reward or other incentives offered by the prize and the definition of the prize challenge. The literature offers some insights on the positive effect of those parameters on innovation. Larger prize incentives are likely to attract more entrants and induce more intense competitions. Larger rewards also increase the chances of engaging unconventional entrants that bring fresh and more creative ideas if the definition of the prize challenge is sufficiently open-ended to allow alternative approaches to solve technological problems. Depending on the characteristics of the technological field, prize technologies might also be introduced to markets or used for alternative applications. Although intuition suggests that kind of results if those parameters are properly set, a null relationship between the prize incentives/challenge definition and innovation might be plausible. If the actual incentive to innovate is in the market value of the prize technologies, for example, prize participants might enter competitions to be able to realize such value regardless of the prize designs.

It should be noted that while for many the winning entry has traditionally been the demonstration of the ability of prizes to induce innovation, this investigation has sought to gather evidence of and examine not only the more visible outputs but also other developments of the competition. From this perspective the winning entry in competitions speaks more about the winner's ability than about the real effect of prizes. Distinguishing whether certain effects are truly induced by the prize or by other contextual factors is an intricate task due to the specificity of each prize. Prizes are unique instances given their context or technological field, competition design and prize participants. Our ability to gain a better understanding of the effect of prizes would be limited if no examination of both the dynamics of the competition and its context is performed.

NOTES

1. Overinvestment in innovation from the social welfare point of view may exist with other incentives as well, when for example competing firms try to get ahead of one another's innovation programs (Dasgupta and Stiglitz, 1980).
2. Although this classification allows a more systematic analysis of the motivations, R&D activities, and technology outputs of prize entrants, a word of caution is necessary. Entrants might evolve over time and thus their perceptions and organization change. Entrants might also change as both the competition and themselves gain visibility. They might for example be able to gather new resources including more experienced or skilled members or adapt their form of organization to be able to raise funding and continue in the competition. All these changes suggest that this kind of classification of entrants might not be efficient for the investigation of prize competitions with long lead times. This does not necessarily affect other short-term prizes with less challenging goals.
3. This investigation probed the importance of seven idea sources that are potentially interesting in this context: theoretical knowledge that entrants already had, available commercial products, past projects of the entrant, industry or space agency projects,

designs found in non-aerospace projects, designs of entrants participating in other prizes, and designs of entrants to the same prize. The investigation was also open to discover other unanticipated idea sources. Moreover, this investigation probed the importance of seven design criteria: technical simplicity, project cost, market value, novelty, reliability, environmental impact, and standardization (compliance with industry standards). The investigation was also open to discover other unanticipated design criteria.

4. Methodological aspects

4.1 METHODOLOGICAL APPROACH

This investigation approached the study of prizes following three key steps. First it developed a theoretical framework based on existing literature on prizes and innovation and accounts of recent prize experiences. This first step also contributed a new innovation model applied to the investigation of technology prizes. Second it tested and revised the model by analyzing retrospectively two pilot case studies of recent prizes. This step was important to strengthen the descriptive power of the model and refine the operationalization of categories. This process of model development also contributed empirical evidence to further substantiate next research steps and the findings of the main case study. The third step involved the investigation of the main case study, probed hypotheses and elaborated theoretical, policy and methodological implications.

To accomplish its goal, this investigation pursued a field-based, mixed methods case study strategy combined with an iterative process to build explanations about the phenomenon and generate grounded theory (Eisenhardt, 1989; Yin, 2003). Different methods of data gathering such as direct observation, interviews, application of questionnaires and document analysis, as described in the following sections, helped to gather richer and more reliable data. The empirical examination of case studies followed an embedded multiple-case study design (Yin, 2003) combining multiple-units of analysis in two stages (Figure 4.1). There was a first stage with two pilot case studies and a second stage with a main case study. The first stage reduced the risk of focusing on a single case study and serves other purposes as well. It allowed defining more precisely the constructs involved in the model, determining the plausibility of the hypotheses, testing data sources, improving data gathering instruments and contributing insights for the main case study. Multiple embedded units of analysis in both stages provided richer data and more variation to analyze the prize cases.

The analysis strategy has been based on the triangulation of data sources during the interpretation of data with equal weighting for data gathered on embedded cases through different methods (Creswell and Plano Clark, 2007). The analysis followed an interactive approach involving

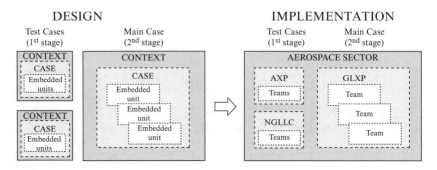

Source: Author's design based on Yin (2003).

Figure 4.1   Case study design and implementation

data reduction, displaying and conclusions (Miles and Huberman, 1994). This approach involved follow-ups, further data gathering and re-coding when there were missing or ambiguous data. Software for data coding and qualitative analysis was used in this process. Case study data include documents, responded questionnaires, transcribed interviews and memos with observation data.

This investigation maintains that we can gain a better understanding of prizes only by examining not just the prize competition as the unit of analysis but also other aspects of the prize phenomenon at the entrant- and context- levels. Therefore the author followed an approach that sought to learn about the prize by looking at how entrants respond to incentives, perform R&D activities and advance technologies, and disentangle the effect of the prize competition from ongoing R&D activities and broader industry and technological trends. This highlights the importance of the historical research perspective when examining cases of past prize competitions. Essential to this investigation has been the consideration of the counter-factual 'Would innovation occur anyway if the prize did not exist?' to increase the external validity of the research findings. For this two additional aspects are considered in case studies: (a) the potential alternative strategies that prize entrants might have been pursued if there was no prize announcement or if prize entrants did not join the competition; and (b) the industry sector/technological field developments that might have occurred if the prize was not announced. Prize entrants have been asked, for example, about their goals and the relationship between the prize and their previous projects to understand their potential alternative strategies if the prize was not announced. Experts were also asked to assess these aspects from the industry context analysis standpoint.

4.2 A MODEL OF INNOVATION APPLIED TO PRIZES

4.2.1 Prior Work

To the author's knowledge no framework or model has been previously offered by the scholarly literature to undertake case study research with a focus on innovation prizes. The academic literature has mostly contributed economic models in which a prize sponsor offers a unique monetary reward (the cash purse) to induce either increasing R&D activity in a specific technological field or the production of a single innovation. These models have generally assumed that innovations are ultimately placed in the public domain and thus the prize winner cannot reap the benefits of monopoly pricing related with the commercialization of the technologies. The work by Wright (1983) is among the earliest applications of formal modeling techniques to compare incentive mechanisms, including prizes. Wright explored the theoretical optimal application of patents, prizes and direct research contracting to induce innovative activity from the standpoint of a social welfare-maximizing administrator and with many researchers targeting the same innovation. In his work the optimal choice between those mechanisms is based on terms of the probability of success of the research projects and the elasticity of supply of research. Several other works have built upon Wright's research to compare the effectiveness of alternative incentive schemes. de Laat (1997) for example investigated whether the results of Wright's model hold under less competitive R&D processes by looking at the case of one innovator that is a technological leader and focusing on the comparison between prizes and patents. Shavell and van Ypersele (1999) also compared the patent system (which incentivizes innovators offering monopoly profits) with a rewards system (which incentivizes research investment through a monetary reward). In their theoretical model there is a single potential innovator with private information about the demand for the innovation and a government with knowledge about the probability distribution of demand curves. Newell and Wilson (2005) analyzed the use of prizes to induce innovations for climate change mitigation and used a simplified economic model that assumes that prize entrants and the sponsor share information about each other's costs, benefits and probability of success. Formal economic modeling of prizes has also been applied to the examination of optimal designs. Taylor (1995) for example introduced a model in which identical risk-neutral research firms, with no capital constraint and private information about the value of the innovations, decide whether to pay an entry fee to participate in a contest to produce an innovation of the highest value for the sponsor and win the prize money. Moldovanu and Sela (2001) used another illustrative model of multi-prize contests where risk-neutral players have private information about their abilities and the number and cost functions of the contestants affect the configuration and size of the prize values that maximize the expected sum of efforts.

There are on the other hand only a number of empirical works that investigated prizes. A notable example is the study of Brunt et al. (2008) which looked at prizes awarded by the Royal Agricultural Society of England (RASE) between 1839 and 1939 to determine whether prizes induce innovation or not. These scholars used econometric models (negative binomial regressions) that look at the entry effect of prizes and correlated patent activity and drew on datasets with prize entrants, rewards and patents data for a time period in which 98 RASE exhibition shows were held. Interestingly their models incorporate the type of reward (i.e. monetary or others) as an independent variable that influences the incentive power of the prize. Another interesting work by Davis and Davis (2004) followed a case study approach to investigate the role and incentive effects of prizes in three 20th century innovations (motorized flight, human-powered flight and energy efficient refrigerators) where prizes were considered to play a significant role in technology development. These scholars examined qualitative, historical evidence to determine how prize designs affected contest outcomes. Although they did not put forward a formal prize model, they examined their cases with guidance of five issues generally addressed by the literature: the size of the prize reward, the potential duplication of R&D efforts, spillovers and reputational gains, sequential innovations and the co-existence of prizes with patents and the firm strategic choice. Another relevant example is the work by Saar (2006) who also drew on case study research but from a different perspective: he investigated the reasons why prizes are rarely used as innovation mechanisms by addressing, among other aspects, the effect of prizes on technological innovation. He studied five prizes offered since the 18th century, namely the Longitude Prize, the Alkali Prize, the Orteig Prize, the Ansari X Prize and the Windows-on-a-Mac Prize. Saar's analysis was structured to answer ten questions that look at diverse aspects of the prize phenomenon, including prize purpose, prize sponsor, prize financing, entry rules, type of prize, related markets, prize offer length, funding of R&D activities, characteristics of the winning solution, publicity of the prize and co-existing prizes. Saar noted among the most important limitations to pursuing this kind of case studies the lack of reliable data, particularly for historical prizes.

Although the investigation presented in this book differs significantly in its research design from previous works, it still benefits from insights that emerge from those other investigations. First of all the number and importance of the conditions under which prizes are demonstrated to be effective (at least theoretically) suggest that new innovation models to investigate prizes should consider a wider range of factors to increase their explanatory power. Those factors are not only at the prize-level but also at the entrant-level (such as attributes of the innovators) and context-level (such as the characteristics and dynamics of the technological field in which the competition is held). To better understand the prize phenomenon, prize models also need to: contemplate a range of possible, more realistic situations such as multiple innovators that might have diverse goals and be involved in the production of more than one innovation; consider the prize incentive in the context of other more widely used traditional incentives such as contracts and grants, to which innovators also have access; and consider that innovators' decisions are not limited to the assessment of information on the costs, benefits and probability of success of research projects but also other aspects related with their personal or organizational strategies and goals. A better understanding of prizes in the context of longer-term R&D activities in a given technological field also requires considering pre- and post-prize activities of the performers and the evolution of such a broader context. In other words, prize research has to consider the sequential and cumulative nature of innovation (Green and Scotchmer, 1995) and examine ongoing industry R&D efforts and the after-market value of the innovations achieved in competitions.

From a methodological standpoint the lack of systematic documentation of a number of aspects related with prize competitions (such as the perception of incentives by prize entrants, R&D activities and/or technology achievements) demands research models that allow empirical investigation based on mixed data sources. In particular prize research requires introducing measures to examine and compare prize technologies and better understand the relationship of those technologies with others that may have been under development before the prize announcement. Intuitively we consider other intermediate outputs of prizes and not only the winning entry, to judge their effect on innovation. The opportunity to observe an ongoing prize competition and activities of entrants is valuable in that regard.

4.2.2 The Model

This research proposes a model of innovation applied to prizes that considers the prize competition as the unit of analysis and contemplates other factors that are internal or external to the competition (Figure 4.2). This model assumes that at the core of each prize competition there is a

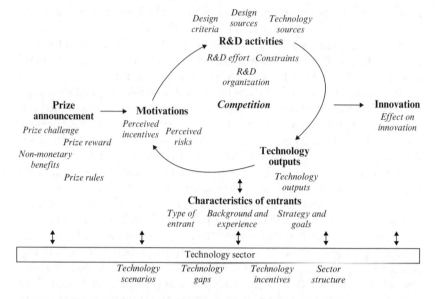

Source: Author's development based on the literature cited in the text.

*Figure 4.2   Innovation model applied to the investigation of innovation
          prizes*

particular pattern that makes prizes unique instances of innovation and
encompasses diverse motivations, R&D activities and technology outputs
of the prize entrants and is influenced by the characteristics of the prize
entrants and the dynamics and structure of the prize context. To expand
its explanatory power the process of model building draws primarily on
the prize literature and considerations made in previous sections. The
model departs from the typical econometric approaches and includes a
comprehensive set of factors that possibly influence the innovation effect
and development of prizes. It also seeks to facilitate prize case study
research that draws on multiple types of data sources.

This model was developed following four main steps. First there is the
identification of dimensions of the study of prizes which involved the analysis of other scholarly works, accounts of historic prizes and analyses and
reports by prize advocates and commentators. In total 40 different knowledge sources published between 1862 and 2009 were analyzed, including
13 scholarly journal publications and 14 working and conference papers.[1]
This initial analysis identified six dimensions of the study of prizes to
outline a research model with its constructs and relationships. These

dimensions comprise prize announcement, motivations of entrants, prize R&D activities, prize technology outputs, characteristics of entrants and characteristics of the context/technology sector. Second there is the identification of themes and topics through literature coding. This task involved the application of a descriptive approach to data coding (Saldaña, 2009) to the selected literature using commercial data coding and qualitative analysis software. Third there is the re-contextualization of topics and their redefinition into categories. Codified topics were re-contextualized from their contiguity-based context into conceptual- or causal-based context based on both prize and more general innovation literature. This task purposely sought to prevent the creation of 'analytic blinders' (Maxwell and Miller, 2008) that may impede observing connecting relationships in the data. This process yielded a number of research categories that follow either a variable-oriented approach in which categories are concepts linked by relationships (e.g. entrant's characteristics that relate to decisions to enter the competition) or a process-oriented approach in which categories are events or conditions that lead to successive events or conditions in the context of one case (e.g. prize design's features that lead to certain technology outputs) (Miles and Huberman, 1994). Finally there is the operationalization of research categories or variables. The categories were defined to be operationally specific (i.e. categories are defined in terms of measurement) and to present no measurement issues (i.e. to be reliable when measuring the phenomenon using different data sources and collection methods) (Bacharach, 1989).

The final version of the model resulted from several iterations of topic-coding and display-building to organize concepts and relationships. The author sought to balance the number of research categories and the scope of the model to be able to address diverse research questions at different levels of analysis. The connections between categories imply correlational or directional relationships that have been probed in case studies. The connecting strategy was simply based on identifying key relationships suggested in the prize literature and more broadly discussed in the innovation literature. Subsets of categories can be linked to more abstract constructs (e.g. motivation) for analytic purposes. The definition and operationalization of research categories were tested with the study of pilot cases as well. The triangulation of data sources helped to test the validity (i.e. appropriateness of operationalization) and reliability (i.e. measurement issues) in categories and to adjust their definitions. The initial application of this model to pilot case studies also helped to probe the utility or descriptive/explanatory power of the model.

A few considerations applied in the operationalization of categories in this model. This investigation considers that motivations are the force that initiates and maintains goal-directed performance and can explain the decisions that convert intent into action (Clark, 2003). To find an explanation of the decisions to participate in prizes, this investigation probed the importance of a set of pre-defined motivations (classified into Very Important/Important/Somewhat important/Not important at all) and allowed research participants to provide additional reasons to participate in questionnaires and interviews. Since the beliefs and ultimate goals of the entrants are likely to moderate or intensify their perception of the incentives offered by prizes and related technologies, this investigation was open to explore explanations to prize participation that consider boundedly rational, complex individuals and organizations that are not mere profit-maximizing entities.

In the assessment of R&D activities and technology outputs the purpose has been to offer insights on the potential contribution of the prizes to space developments and not their technical analysis. Moreover, this assessment is limited to mission approaches and sub-systems and does not include more specific analysis of parts or components. To assess the degree of novelty of the GLXP technologies in particular, entrants were asked to report whether they build completely new systems and components (i.e. from scratch) or acquire, adapt or copy existing technologies. In particular questionnaires asked this in relation with seven main systems of a typical GLXP mission: lunar lander, lunar rover, photo/video system, control/ navigation hardware and software, Earth-to-Moon vehicle, communications between Moon and Earth and ground support systems. Further analysis allowed an assessment of their novelty and potential to become truly innovations as a result of their commercialization or introduction for the entrants' own use.

Finally, these prizes are embedded in a mostly government-led sector and technological development pathways are therefore highly dependent on priorities given by government space programs to certain areas of space development. In this case technology roadmaps prepared by agencies such as NASA are good sources to assess the industry's technological pathways and expected developments. These roadmaps do not make certain what technologies will ultimately be developed, but put forward priorities that will be addressed by agencies and their contractors to accomplish programmed missions. We might expect that, notwithstanding the announcement of the prize, industry efforts are likely to focus on the development of technologies for those future space missions. Other sources such as the scientific and specialized industry literature also helped to understand space developments, as they generally address emerging technologies and new projects undertaken in this sector.

4.3   CASE STUDIES

This research investigates three cases of recent technology prizes in the aerospace sector (Table 4.1). The main case study is the Google Lunar X Prize (GLXP) for robotic exploration of the Moon. The two pilot case studies are the Ansari X Prize (AXP) for suborbital manned flight and the Northrop Grumman Lunar Lander Challenge (NGLLC) for the development of vertical take-off/landing vehicles.

The GLXP[2] is a $30 million multi-year global competition organized by the XPF and sponsored by Google Inc. It was announced on September 2007 and has not found a winner yet. The GLXP requires participants to land a robot on the Moon among other secondary goals by December 2015. Thirty-five international teams entered the competition and more than 40 countries have been involved. This research investigates the competition as unit of analysis, the space sector as its context and 17 teams as embedded case studies. Four reasons made this prize an obvious choice for a main case study. First, the GLXP is an ongoing competition which allows gathering real-time data and observing R&D activities and other aspects of the participation of prize entrants. Second, in addition to the opportunities to gather original data, this is the most documented prize competition in history, with diverse, multiple data sources available. Third, the GLXP prize challenge involves technology development in strategic areas for S&T and innovation policy in many countries. Fourth, the widespread activity induced globally by the GLXP is an opportunity to observe the extent of the effect of prizes and a more diverse range of entrants. In this case the embedded units of analysis comprise those teams that responded to a questionnaire applied to all GLXP teams. To perform a more in-depth examination of prize entrants, the author also selected interview teams based on their willingness to share information about their prize activities, their availability for in-person or phone interviews, the authorization given to observe their activities and facilities, and a more efficient use of the funding available for this research (based on criteria such as proximity of team or cost of travel). The author also sought to increase the diversity of respondents by including both USA and foreign teams and different types of entities and organizational forms.

The AXP[3] is a $10 million prize offered in 1996 to the first non-government organization to launch a reusable manned spacecraft into space twice within two weeks to a minimum altitude of 100 kilometers. It engaged 26 teams from seven countries. The USA aircraft design company Scaled Composites won this prize in 2004. This prize was privately funded and inspired by the early 20th century Orteig Prize for the first nonstop

Table 4.1   *Main characteristics of prizes investigated in this book*

	Ansari X Prize (1996–2004)	Northrop Grumman Lunar Lander Challenge (2006–09)	Google Lunar X Prize (2007–present)
Prize challenge	First non-governmental organization to build and launch a reusable manned space-craft into space twice within two weeks	Build and fly a reusable, rocket-powered vehicle simulating a flight on the moon within pre-specified timeframe and performance and in a designated location	First to land a spacecraft on the Moon, traverse 500 meters and send back high-definition video footage
Prize type	First-to-achieve, winner-takes-all, medium- or long-term competition	Best-in-class, multi-prize, multi-year competition with purse rollover	First-to-achieve, multi-prize, medium- or long-term competition
Prize purse	$10 million	Level I: $350 000 for 1st place; $150 000 for 2nd place Level II: $1 million for 1st place; $500 000 for 2nd place	Grand prize: $20 million 2nd place prize: $5 million Bonus prizes: $5 million
Sponsor/ manager	XPF (sponsor and manager) through insurance policy with funding from the Ansari family	NASA and Northrop Grumman Corp. (sponsors) / XPF (manager)	Google Inc. (sponsor) / XPF (manager)
Prize entrants	26 teams from seven countries	12 USA teams	35 teams from 17 countries (6 already withdrawn)
Prize winners	Scaled Composites, from Mojave, California	NGLLC 2006 and 2007: No winners NGLLC 2008: Armadillo Aerospace from Rockwall, Texas: Level I (1st place)	No winner yet

Table 4.1 (continued)

Ansari X Prize (1996–2004)	Northrop Grumman Lunar Lander Challenge (2006–09)	Google Lunar X Prize (2007–present)
	NGLLC 2009: Masten Space Systems from Mojave, California: Level I (2nd place); Level II (1st place) Armadillo Aerospace from Rockwall, Texas: Level II (2nd place)	

Source: Diverse sources described in the following chapters.

transatlantic flight between New York and Paris. The AXP was the first prize program administered by the XPF.

The NGLLC[4] is a multi-year competition held between 2006 and 2009 as part of NASA's Centennial Challenges program, which comprises about a dozen different prizes. Twelve independent small USA teams participated in four years of competition. The NGLLC offered a total of $2 million in cash prizes for the first and second best-performing teams. To win, the teams had to build and fly a vertical take-off and landing rocket-powered aircraft within minimum pre-specified standards of efficiency and under conditions that simulate the same flight on the moon. This program had two competition levels (I and II) with different degrees of difficulty (II being the more difficult). The prize money rolled over to the next year when no entries qualified. Masten Space Systems and Armadillo Aerospace, two aerospace startups, won different levels of this prize in 2008 and 2009 and shared the total prize money.

The criteria to select the AXP and the NGLLC were the similitude of industry sector with the main case, which facilitates the observation of teams that participated in more than one prize, the simplification of the investigation by studying only one context without significantly affecting the results of the project, and the elaboration of implications specific to the use of prizes in a strategic sector such as aerospace. The entrants studied as embedded cases were selected based on the availability of secondary data on their prize participation.

4.4   DATA AND DATA GATHERING

4.4.1   Pilot Case Studies

The investigation of the AXP and the NGLLC draws mainly on the analysis of documentary sources. Seven embedded cases (out of 26 prize entrants) are examined for the AXP and five embedded cases (out of 12 prize entrants) are examined for the NGLLC. Eighty-seven primary and secondary data sources (McDowell, 2002; Danto, 2008) have been codified to gather case study data including research articles, books, websites and other Internet content. Prize case studies of competitions held in the past require the researcher to disentangle all the historical and special interest factors that might affect the objectivity of the documentary sources. Data sources were checked for external validity (i.e. authenticity) and internal reliability (i.e. credibility and biases) (Danto, 2008). Comparing and contrasting data sources were very important to tease out common components and attributes, interpret and reconstruct the story of the prize competitions and the participation of entrants. The triangulation of data sources also helps to increase the internal validity of the research. Primary data sources provide data directly from prize entrants in the form of team profiles, blog posts, forum comments and other online content created by team members. Secondary data sources comprise all other sources including but not limited to the prize sponsor's reports and media coverage. Both competitions were held in a time period that coincides with the widespread growth of the Internet and that offered a platform for the publication of information about the competitions and participant teams. Online tools such as the Wayback Machine also allow gathering data from web pages that existed only when these competitions were in progress.[5] Moreover, prize sponsors have been interested in disseminating information and progress updates on these competitions to attract the attention of the general public, which increased the amount of available data. Further data on the AXP and the NGLLC were gathered in unstructured, open-ended interviews with prize experts. Mr. Gregg Maryniak (Vice President of Aerospace Science, St. Louis Science Center and X Prize Foundation's Advisor) managed diverse aspects of the XPF in the 1990s and was interviewed by phone to learn more about the AXP. Mr. Ken Davidian (Director of Research at the FAA Office of Commercial Space Transportation – AST) is former NASA's manager for the Centennial Challenges program and was interviewed in-person to learn more about the NGLLC. The knowledge and expertise of these interviewees also allowed learning more about other prize cases including the GLXP.

4.4.2 Main Case Study

The study of the GLXP draws upon data collected at the entrant, competition and context levels. Team-level data sources comprise questionnaires to teams, interviews with team leaders and members, site visits and documentary sources (e.g. team profiles posted on official GLXP website, team websites). Prize-level data sources include an interview with the XPF's GLXP Prize Manager and documentary sources (e.g. official GLXP website, official press releases). Context-level data sources comprise interviews with industry experts (N=3 experts) and documentary sources (e.g. journal articles, industry reports, media articles).

Questionnaires to GLXP prize entrants were mailed to leaders of 23 GLXP teams between February 2010 and September 2010.[6] A total of 17 teams responded to the questionnaire within that time frame. This response rate allowed gathering data from both USA and foreign teams in proportions similar to the overall participation of USA and foreign teams throughout the competition. One of the participating teams was already withdrawn when responding to the questionnaire (a special version of the questionnaire was mailed with similar questions). Three other participating teams withdrew from the competition after the data gathering process finished. The rest of the teams are still in competition. On the other hand the other six teams that did not participate in this investigation either did not respond to the request to fill out questionnaires or responded that they were too busy to participate in the study. On these teams and those that entered the competition after the data gathering time period, the author was only able to gather data from documentary sources which include team websites, profiles on the official GLXP website and other Internet media content. The questionnaire was organized in five sections and asked team leaders about the motivation, R&D activities, technology outputs and members of their teams. Most of the creativity and innovation literature draws upon self-reporting methods when studying team process variables (Hulsheger et al., 2009). Innovation research based on self-report data such as questionnaires, however, may suffer the problems of susceptibility of response biases and potential overestimation of effect sizes. This investigation sought to overcome such potential problems with further data gathering and triangulation of multiple data sources.

This investigation considered that the analysis of questionnaire data might suffer from nonresponse bias. Nonresponse bias results from individuals or organizations who respond to questionnaires being different from individuals or organizations who do not respond, in a way relevant to the study (Dillman, 2000). There may be different sources of nonresponse. Refusal and 'unlocated' apply to this type of research (Daniel, 1975; Hawkins, 1975). Refusals to respond may depend on researcher–subject rapport, the quality of the questionnaire and the nature of the inquiry (which might include confidential or sensitive information in this case). Unlocated situations may occur when no contact data are available. To handle nonresponse and improve the questionnaire's clarity and acceptability, the questionnaire design was pre-tested for question comprehension, ability to answer questions without error, selection of responses and reactions to sensitive questions. This pre-test was performed with the help of three anonymous colleagues and four graduate students from the Satellite Communication and Navigation Systems class of Fall 2007 at the Georgia Institute of Technology.[7] To increase the response rate to questionnaires this investigation had collaboration from the XPF to deliver envelopes with questionnaire/letter packages to team leaders at the 3rd annual GLXP Summit. The XPF also provided contact data of teams for follow-ups when team leaders authorized that. The investigation also used the personalization technique in e-mail follow-ups which involves giving the researcher's attention to individual team leaders (Dillman, 2000).

The interviews with team leaders and members addressed most of the topics of the questionnaire and allowed discussing more in-depth specific aspects and obtaining clarification of responses given in questionnaires. These interviews were semi-structured and open-ended, and were conducted in-person at the team workplaces in most cases. Seven teams accepted interviews with their team leaders and in some cases with team members as well. One interview was conducted by phone and another one by sending questions and receiving responses by e-mail. When given the proper authorization, the researcher visited the workplaces of the teams to observe how the teams organize their activities. Five teams were visited including eight different facilities or workplaces.

The analysis of team-level data does not seek to compare teams based on their performance but to discover patterns that help to better understand how motivations, R&D efforts and technology development occur in the context of prizes. Since the GLXP is an ongoing competition, the data that could reveal competitive positions or strategies of teams were anonymized or deliberately excluded from this analysis. The perception of incentives and the self-assessment of the outputs of the teams and their innovativeness date to the moment in which each questionnaire was answered. The author sought to consider the external factors that could affect perceptions and assessments during the time period in which the questionnaires were answered.

To gather data on the GLXP competition the investigation used three main data sources. First, the researcher conducted an unstructured, open-ended interview with Mr. William Pomerantz, the XPF's Director for Space Prizes, in person which helped to better understand diverse aspects of the prize and its context. The researcher also attended the 4th annual GLXP Summit held in the Isle of Man (UK) in October 2010. At this event representatives of 13 teams presented updates on their projects and discussed different aspects of organization of the competition with the organizers. This opportunity was used to observe the interaction between team members and between the teams and the prize sponsors, and to speak informally with some team members to gather general impressions about the Summit and the competition. Finally this investigation drew on documentary sources, including the official website of the GLXP, the websites of the teams and diverse content published on the Internet by the specialized media.

To investigate the GLXP context this investigation includes insights from industry experts and literature analysis. The researcher conducted unstructured, open-ended phone interviews with three experts with significant experience from space agencies, traditional space industry and the emerging new space sector. The interviewees include Mr. Dennis Stone, Assistant Manager for Commercial Space Development in NASA's Commercial Crew & Cargo Program at the Johnson Space Center; Mr. Jeff Greason, founder and President of XCOR Aerospace and founder of the Commercial Spaceflight Federation; and G. Thomas Marsh, former Executive Vice President of Lockheed Martin Space Systems Co. These experts contributed opinions and points of view in relation to four main aspects of the context of the GLXP: expected sector scenarios, technology gaps to achieve prize targets, industry structure and technology-related incentives linked to the GLXP and other space prizes. The contribution of the experts to each of these aspects varies. Interview responses could not be more precise without further analysis on the experts' part, particularly in the case of questions that concern expected technology scenarios. These experts still provided significant insights that allow a better understanding of the context in which the GLXP is held. The opinions of these experts do not represent those of their organizations.

NOTES

1. All references to this literature are cited in this book.
2. Google Lunar X Prize's official website: http://www.googlelunarxprize.org/
3. Ansari X Prize's official website: http://space.xprize.org/ansari-x-prize
4. Northrop Grumman Lunar Lander Challenge's official website: http://space.xprize.org/lunar-lander-challenge

5. The Wayback Machine website can be accessed at: http://www.archive.org/web/web.php
6. Only one team, whose leaders were involved in a legal process at that moment, was not surveyed.
7. These students were familiar with the GLXP as they worked on class projects aimed at designing a spacecraft that could potentially participate in the GLXP that was just announced.

5. A first approach: the Ansari X Prize and the Northrop Grumman Lunar Lander Challenge

5.1 THE ANSARI X PRIZE

5.1.1 The Prize

The AXP was announced by the XPF in 1996. It offered a $10 million cash purse for the first non-governmental organization to build and launch a reusable manned spacecraft into space twice within two weeks by 1 January 2005. This prize was privately funded and inspired by the early 20th century Orteig Prize for the first nonstop transatlantic flight between New York and Paris. Twenty-six teams from seven different countries entered this prize. The competition was won in 2004 by Scaled Composites, a USA aircraft design company. The winning flights are considered the first privately funded human spaceflights in history.

This was the first prize program administered by the newly created XPF, an educational, non-profit corporation established in 1994 to inspire private, entrepreneurial advancements in space travel. It is also one of the first and most popular modern innovation prizes launched since the 1990s. Its purpose was to demonstrate the feasibility of private space flight, change existing public opinion about the private space industry's capabilities and generate concrete business opportunities for commercial space tourism (Maryniak, 2010). The prize challenge involved building and flying a manned vehicle to a suborbital space and having mostly privately funded projects. Although the idea of space tourism was not new at that time, the AXP defined the private space flight problem in concrete, measurable terms for the first time. In particular the rules established these four key requirements:

1. *Space flight*
    The spacecraft had to be able to reach an altitude of 100 kilometers, the minimum altitude to be considered a space flight.

2. *Crew capability*
   The spacecraft had to be able to carry the pilot plus equivalent capacity for two passengers.
3. *Re-Flight*
   The same vehicle had to complete two flights within two weeks.
4. *Reusability*
   No more than 10 per cent of the vehicle's first-flight non-propellant
   mass could be replaced between the two flights.

The teams were allowed to use any technological approach to accomplish this feat (the rules mention for example tow vehicles, balloons and
descent ballutes, among other examples) but they were required to fund
90 per cent or more of their projects with private sources. The prize rules
allowed teams to retain all the IP and commercial rights related to their
technologies, vehicles and services.

Unless otherwise indicated the analysis and findings reported in this
section are based on seven embedded cases or teams: Scaled Composites
(USA), Armadillo Aerospace (USA), Advent Launch Services (USA),
ARCA (Romania), Da Vinci Project (Canada), PanAero (USA) and
Starchaser Industries (England) (Table 5.1). A complete list of AXP
entrants has been included in the Appendix (Table A.1).[1]

5.1.2   The Context

Space tourism was not a new idea when the AXP was announced.
Suborbital pleasure trips began to receive serious consideration in the mid-
1980s but the cost of space travel was still the main barrier (Collins and
Ashford, 1986). As early as 1994 the largest aerospace companies in the
USA were aware of the potential space tourism market but also of the need
to increase safety and reliability and reduce per passenger costs significantly (Boeing et al., 1994). The potential opportunities in this market were
also clearly appreciated in 2001 when Dennis Tito became the first paying
space tourist. His flight to the International Space Station (ISS) in the low
Earth orbit had a $20 million price tag and demonstrated the possibilities
for private space travel. A few other wealthy tourists have repeated such
trips and many others have been willing to, at least to the suborbital space.
In 2002 a study by Zogby International, a USA market research firm,
found that 19 per cent of the individuals surveyed in the USA were willing
to pay $100000 for a 15-minute trip into suborbital space (Byko, 2004).

There was however no established provider of suborbital space flights
as of 1996. Large companies in the aerospace and aviation business (such
as Lockheed Martin Corp. and The Boeing Company) demonstrated no

Table 5.1 Summary of data for embedded cases in the AXP

				Selected teams			
	Scaled Composites	Armadillo Aerospace	Advent Launch Services	ARCA	Da Vinci Project	PanAero	Starchaser Industries
Team description	Pre-existing aircraft design firm	Newly created, independent R&D team	Employee-owned corporation	Non-profit org. pursuing space activities, created by students	Independent R&D team	Newly created aerospace engineering company	Space research foundation later incorporated as company to enter the prize
Experience / Background	Vast experience in innovative aircraft design	Diverse non-aerospace backgrounds	Extensive experience from NASA	Aeronautical engineering students	Aerospace support systems background	Extensive NASA experience	Aerospace experience
Created	1982	2000	1996	1999	N/A	1997	1998
# members	135	6	12 (~100 volunteers)	8	14 (~500 volunteers)	9	35
Based in	Mojave, CA, USA	Mesquite, TX, USA	Houston, TX, USA	Ramnicu Valcea, Romania	Toronto, Canada	Fairfax, VA, USA	Cheshire, England

Note: N/A indicates data not available.

Source: Author's analysis based on data sources described in the text.

interest in this competition when it was announced, something that the XPF sought purposely with its prize design (Diamandis, 2004; Maryniak, 2010). A few other small companies were also pursuing similar suborbital flight targets. An interesting example is XCOR Aerospace, a USA private rocket engine and spaceflight development company founded in 1999. While the company had projects similar to the challenge posed by the AXP, it did not enter the competition because the prize was perceived as a reward for speed of development rather than for the commercial merit of the spacecraft technologies (Greason, 2010).

The competition received extensive media coverage due to its implications for space exploration (XPF, 2007). At the time it was won, this competition received more than five billion media impressions and was telecast and webcast to a global audience with the support of NASA, America Online, the Discovery Channel and other media outlets (Maryniak, 2005). As of 2011 commercial tourism space flights are not more frequent than those early experiences of ten years ago. Industry experts suggest that this market might be sizable, but only more affordable and safe technological solutions are likely to have commercial viability (Marsh, 2011). Interestingly no flight permit had ever been given by the US Federal Aviation Administration (FAA) for private human spaceflight before the winning attempts of 2004.

5.1.3   Prize Entrants

Prize registration was open to any kind of private team (including international teams) with the condition that its spacecraft had to be mostly privately financed and built. Ultimately 26 official teams from seven countries entered the competition. The XPF however received many more inquiries from potential entrants interested in this prize (Maryniak, 2010). The data suggest that at least 18 out of 26 prize entrants were unconventional to the space sector, including pre-existing companies that re-directed their activities, new startups and independent R&D groups. The members of teams such as Advent Launch Services, PanAero and Starchaser Industries (among those investigated here) had more extensive space agency/industry experience.[2] The XPF once defined the AXP teams as 'people that would never look at a government contract' (Diamandis, 2004). Large traditional industry players such as Lockheed Martin Corp. and The Boeing Company did not enter the prize but there were a few newly created teams with more significant space agency/industry linkages. The team Advent Launch Services for example was founded by a group of NASA retirees.

Scaled Composites, the winner, is a USA aviation company that has been building innovative aircraft since 1982 (Byko, 2004). Northrop Grumman Corp. had a 40 per cent-stake in this company when it won the AXP. Most of the other teams were created after the year of the prize announcement and, to the author's knowledge, did not have that kind of corporate ownership. There were very small teams of about 10 people or fewer and sometimes they gathered significant volunteer efforts as well. The teams Advent Launch Services and Da Vinci Project for example enrolled up to 100 and 500 volunteers, respectively. Scaled Composites had about 50 full-time equivalent people working on the AXP project (XPF, 2008b). Some kind of engineering experience was a feature of all teams.

Three out of seven teams analyzed here entered the competition in 1996/1997 and four entered after 2000 (overall more than 45 per cent of the AXP teams entered in or after 2000). Scaled Composites entered the competition officially in 2001. The same year the company partnered with Vulcan Inc. (owned by Paul Allen, cofounder of Microsoft) to create Mojave Space Ventures and started the effective development of the winning entry (Discovery Channel, 2005; Linehan, 2008). While some literature points out that Scaled Composites had its own ongoing 'secret space program' since 1993 (Byko, 2004; Linehan, 2008) experts contribute different opinions on whether the technologies under development were actually useful to win this prize (Greason, 2010; Maryniak, 2010; Marsh, 2011).

5.1.4 Motivations

There were diverse motivations to enter the AXP but those related with economic values still predominate in more than half of the teams. The $10 million cash purse is still secondary compared with the potentially sizable market for the technologies involved in the competition. This prize offered several resources to those teams interested in the pursuit of a commercial enterprise. Advent Launch Services and Starchaser Industries considered the cash purse as potential funding to pursue related projects or start a new company. The competition may have also helped teams to increase their reputation to attract investors, a main pursuit of teams such as PanAero. The main reasons for Scaled Composites to enter the AXP were most likely to be related with the potential market value of space flight technologies and the public exposure and reputation the prize could potentially offer to the company's project. The ability to retain the IP rights on their technologies and services may have been decisive as well for those entrants interested in technology commercialization. Teams such as Armadillo Aerospace and Da Vinci Project were particularly incentivized by the opportunity to pursue a challenging goal, and others such as ARCA saw in the AXP an opportunity to accomplish other personal and organizational goals. The prize also represented an opportunity to learn more about aerospace technology development, something that was particularly interesting for teams such as Armadillo Aerospace.

There is no evidence that these teams considered entering the competition as an additional risk to the risk implicit in technology development. Some evidence shows that Scaled Composites and Starchaser Industries particularly refer to using risk management and safety procedures in technology testing. Although the XPF needed until 2002 to secure the money necessary to pay the reward, only one team commented about the risk of not being paid the reward if it achieved the prize challenge.

5.1.5   R&D Activities

Most of the AXP teams presented novel conceptual designs for their spacecraft. They focused however on providing simple and low cost solutions to the prize challenge. Only three of these teams specifically referred to reliability as a design criterion, and safety was explicitly referred to in only two of these seven projects. By the prize rules all teams had to consider the reusability, re-flight and minimum crew capability requirements in their designs. Design ideas came from multiple sources, but most of these teams refer to existing technologies as the predominant source. The designs of Scaled Composites were particularly influenced by previous experience, the X-15 rocket-powered aircraft project (a USA USAAF/ NACA program) and a criterion that sought to shorten development lead times (Discovery Channel, 2005). Da Vinci Project proposed an unorthodox air-based launch scheme using balloons and existing rocket technologies. Other teams such as PanAero expected to use rocket-powered designs based off existing aircrafts.

A few teams concentrated most of the R&D activity in this competition. Only Scaled Composites and Da Vinci Project scheduled an attempt to win the prize and only three teams actually tested spacecraft or scaled-down versions of them (Scaled Composites, Da Vinci Project and Starchaser Industries) (Linehan, 2008). Scaled Composites exerted the greatest R&D effort, estimated at $30 million, with support from investors and access to significant human resources (Linehan, 2008). The effective development lead time of this effort (final design and building of the winning entry) started in 2001 after this team entered the competition officially (Discovery Channel, 2005). Da Vinci Project was also able to raise some funding from investors. American Astronautics Corp. invested $2.5 million. Other R&D efforts were more modest and generally self-funded. Armadillo Aerospace for example, a fully self-funded volunteer team, invested about $1 million.

Teams ARCA and PanAero may have had even smaller budgets (Byko, 2004; Culver et al., 2007). Overall the 26 AXP teams spent more than $100 million in their prize participation (XPF, 2004). Teams such as Da Vinci Project and Advent Launch Services involved many volunteers as well.

These seven teams manufactured, subcontracted and procured commercial off-the-shelf technologies to different extents. Teams such as Armadillo Aerospace and ARCA drew particularly on in-house manufacturing. Da Vinci Project and Scaled Composites developed their technologies in-house but subcontracted parts and sourced off-the-shelf technologies. The teams also implemented diverse R&D approaches. Scaled Composites for example applied a fast prototyping approach in which an interchangeable group of engineers and technicians can rapidly define a technical or mathematical problem and apply their expertise quickly to solve it, even on the business side (Kemp, 2007). This team even launched its own procurement competition to source rocket technologies from two other companies. Armadillo Aerospace was a small, entrepreneurial organization characterized by a fast-prototyping approach. Da Vinci Project coordinated a large pool of volunteers and consultants. PanAero involved a partnership of companies pursuing the space tourism market. Some teams drew upon knowledge and advice provided by consultants (e.g. Da Vinci Project) or even family and friends (e.g. Advent Launch Services.) In general the main obstacle for the projects of these teams was the lack of funding, with only Scaled Composites and Armadillo Aerospace specifically referring to the lack of aerospace engineering experience and skills. Regulatory requirements were a barrier common to all teams. No team reported facing constraining prize rules.

5.1.6 Technology Outputs

While the AXP did not directly specify the characteristics of the technologies that had to be used to accomplish the prize challenge, winning the prize required a creative approach to building and operating a space vehicle with a relatively low budget and certain minimum capability requirements. The openness of the prize rules in terms of technology requirements resulted in the presentation of very diverse conceptual designs, but only a few teams ultimately entered in development and testing phases. The spacecraft designed by Scaled Composites has been considered a significant innovation in the field. This winning entry combined the White Knight turbojet aircraft with a SpaceShipOne spacecraft attached below, both built by Scaled Composites (Figure 5.1). The novel features of that spacecraft comprise a hybrid design configuration that uses a pivoting-wing system, a patented hybrid rocket motor configuration, an air-launch

Source: Photo courtesy of Mojave Aerospace Ventures, LLC.

*Figure 5.1   Scaled Composites' White Knight turbojet aircraft with
            SpaceShipOne spacecraft attached underneath*

system for piloted spaceflight and the introduction of new materials to
space projects (Boyle, 2004). Other teams also contributed significant
outputs. Da Vinci Project tested a scaled version of its alternative launch
system. Team ARCA introduced the first monopropellant, composite
materials, fully reusable rocket engine. Armadillo Aerospace performed
multiple tests introducing novel operational features and sophisticated
computer controls in its spacecraft.

None of the spacecraft designed and developed for the AXP were
flight-ready for commercialization of space tourism services by the time
this prize was won. Scaled Composites' spacecraft demonstrated its
capabilities but required further development to operate under regular
service conditions. On the other hand only three out of these seven teams
revealed the emerging space tourism market as their main target when
entering the competition, and a further three teams explicitly excluded
commercialization of technologies as the purpose of their participation.
Scaled Composites successfully commercialized its technologies. In 2004
an agreement between the company and Virgin Galactic resulted in a $250
million contract to deliver a fleet of seven spacecraft to offer suborbital

travel services (Linehan, 2008). Despite not having a for-profit orientation, the team Da Vinci Project also partnered with a company to eventually enter the space tourism market. Starchaser Industries pre-sold seats in
its space flight to be able to fund its project. Other teams with commercial
orientation such as Advent Launch Services were not able to commercialize their technologies during the competition.

5.2 THE NORTHROP GRUMMAN LUNAR LANDER CHALLENGE

5.2.1 The Prize

The NGLLC was a multi-year competition held between 2006 and 2009.
It was part of NASA's Centennial Challenges program which comprises
about a dozen different prize competitions. The NGLLC involved only
twelve independent, small USA teams in four years of competition but
many more demonstrated interest in participation. In the first year of
competition, for example, up to 45 other potential entrants demonstrated
interest in this competition (Pomerantz, 2006). The NGLLC offered a
total of $2 million in cash prizes for the first and second best-performing
teams. To win, teams had to build and fly a vertical take-off and landing
(VTOL) rocket-powered vehicle within minimum, pre-specified standards
of efficiency and under conditions that simulate the same flight on the
Moon. The goal of the NGLLC was to accelerate commercial technological developments that would have direct application to NASA's space
exploration goals (including the development of a new generation of
lunar landers) and the commercial launch procurement market (Northrop
Grumman, 2007). NASA, Northrop Grumman Corporation and the
XPF partnered to offer this prize. The XPF managed the competitions at
no cost to NASA. This agency contributed the cash purse and Northrop
Grumman funded part of the costs of operation of the program.

   This program had two competition levels with different degrees of difficulty, I and II, the latter being the more difficult. The prize money rolled
over to the next year when no entries qualified in each level. In 2006 and
2007 the prize-winning attempts of all the teams took place at a sponsor-
organized public event. The same format was used in 2008 but the event
was not open to the public. In 2009 the teams were allowed to designate
their preferred site and date to attempt their flights. Masten Space Systems
and Armadillo Aerospace, two aerospace startups, shared the prize money
for Level I in 2008 and Level II in 2009. In 2008 Armadillo Aerospace
won the first place for Level I ($350000). In 2009 Masten Space Systems

won the second place for Level I ($150 000) and the first place for Level II ($1 million), and Armadillo Aerospace won the second place for Level II ($500 000). By the time this prize was launched, experts considered these two teams the favorites to win the prize (Greason, 2010).

Unless otherwise indicated, the analysis and findings presented in this section are based on five embedded cases or teams: Armadillo Aerospace, Masten Space Systems, BonNova, High Expectations Rocketry and Unreasonable Rocket (Table 5.2). A complete list of NGLLC entrants has been included in the Appendix (Table A.2).

5.2.2   The Context

The first R&D activities specifically aimed at building lunar landers were performed in the 1960s when NASA procured technologies that ultimately were used in the Apollo missions to the Moon. This technology sector has been primarily driven by programs of NASA and other USA and foreign government agencies. Large corporations have dominated most of the private space technology market until the recent emergence of new aerospace development startups in the 1990s. Northrop Grumman for instance is a $30 billion global defense and technology corporation that has been NASA's prime contractor in the development of several technologies linked to lunar programs since the 1960s, including the design and production of the Apollo lunar modules.[3] Interestingly no significant developments were made in this kind of technology in the time period between the Apollo program and some years before the prize announcement. The $60 million program to develop the Delta Clipper DC-X experimental vehicle in the 1990s was possibly the most recent antecedent in the development of VTOL vehicles (Pomerantz, 2007). That program was a technology demonstration project first led by the US Department of Defense and then by NASA, with participation of McDonnell Douglas Corporation. Interestingly Scaled Composites, the winner of the AXP, contributed technologies to the program as well. Other new space companies such as Blue Origin have developed similar VTOL technologies. Blue Origin's New Shepard spacecraft in particular is inspired by the old NASA Delta Clipper DC-X concept (NASA, 2010a).

5.2.3   Prize Entrants

The NGLLC enrolled only 12 unique teams in four years of competition: four in 2006, eight in 2007, nine in 2008 and three in 2009. Some teams participated more than once, in different years. Only USA teams were eligible to enter this competition. USA government organizations, organizations

Table 5.2 Summary of data for embedded cases in the NGLLC

			Selected teams		
	Armadillo Aerospace	Masten Space Systems	BonNova	High Expectations Rocketry	Unreasonable Rocket
Team description	Independent R&D team	Small startup, rocketry and propulsion company	Small design company	Small engineering research group	Father and son amateur team
Experience / Background	Gained specific engineering experience only after joining the AXP and several years of R&D	Only one member with aerospace experience and the rest with Internet technology background	Leader participated in design of AXP's winning entry, yet very diverse backgrounds the rest	Diverse engineering experience and background	Engineering background yet no experience with rocket engines until competition
Created	2000	2004	N/A	N/A	2007
Years of participation in NGLLC	2006, 2007, 2008 (1st place Level I), 2009 (2nd place Level II)	2007, 2009 (2nd place Level I, 1st place Level II)	2007, 2008, 2009 (withdrawn)	2008	2007, 2008, 2009
# members	8	5	6	4	4
Location	Mesquite, TX	Mojave, CA	Tarzana, CA	Moscow, ID	Solana Beach, CA

Note: N/A indicates data not available.

Source: Author's analysis based on data sources described in the text.

principally or substantially funded by the federal government and government employees were not eligible. Most of the prize entrants were unconventional teams organized as independent R&D groups that were created to compete in this or previous prizes. The five teams considered here were generally very small (between five and nine people) and self-funded with some kind of (diverse) engineering experience, including some team members with amateur or professional rocketry experience. Three of them were based in California. Armadillo Aerospace for example entered the AXP a few years earlier and consolidated as an aerospace startup by the time it entered the NGLLC. This team did not have paid employees when it entered this competition in 2006, and worked only two days a week on its project (Pomerantz, 2006). High Expectations Rocketry was another small engineering and research group with some rocketry, engineering and software experience that entered in the third year of competition. BonNova, also a multi-year participant, was a small design company with some aerospace experience. Its team leader was part of the team that developed the winning entry of the AXP. Unreasonable Rocket, another multi-year participant, was a small amateur team formed by a father and son. Masten Space Systems was organized as a small startup, rocketry and propulsion company with six full time employees in 2006. While this was a two-year old company by the time of the prize announcement, it is not the kind of traditional industry player we might expect in this sector.[4]

5.2.4 Motivations

There were diverse motivations to enter the NGLLC but again, monetary motivations seem to predominate. Three out of five analyzed here declared primary interest in the prize money. While the cash purse was very small compared to the cost of related technology development programs, it still was a significant amount of money for these small teams. These teams (particularly Armadillo Aerospace and Masten Space Systems) also sought to build a good reputation and gain the respect of observers that included NASA's and other corporate officials in addition to the general public (XPF, 2007). Although the prize did not represent any commitment from the sponsors to acquire technologies, the idea of potential contracts could not be discarded if one considers the contract-like, narrowly defined technological target and detailed prize technical specifications. Moreover, prize entrants retained the IP rights for their technologies yet were required to negotiate in good faith to provide non-exclusive licenses if NASA was eventually interested in that. A similar number of teams were driven by the challenging goal and the opportunity to participate in a project that team members personally enjoyed. There are in this case also examples of how incentives play a role at the individual team member level. Masten Space Systems for example saw an increase in interest in its employees after it won the prize (Masten Space Systems, 2010), suggesting the opportunity for team members to personally benefit from their involvement.

There is some evidence on the risk perceptions of these teams. They referred to the risk of participation in different terms, including the probability of not being paid the cash purse, destruction of profitable equipment or excessive personal commitment. Masten Space Systems for example declared it had to balance the risks of participating and using the company's equipment with its business risks. In interviews, team BonNova's leader put special emphasis on the personal commitment a prize requires and his considerations before deciding to enter another prize. In one of its blog entries the team Unreasonable Rocket published a stochastic assessment of the risk of participation combined with a cost-benefit analysis.

5.2.5 R&D Activities

These five teams focused on delivering simple and low cost solutions. Team Armadillo Aerospace introduced other non-aerospace design criteria such as modularization and programmability. Masten Space Systems also put special emphasis on reliability and low maintenance, which is associated with commercially oriented designs. Team BonNova also focused on efficiency. Idea sourcing was very diverse across teams. Armadillo Aerospace introduced new designs based on existing technologies. Masten Space Systems drew mostly on own know-how. BonNova also sourced ideas from fiction and other industries as well (a BonNova team member was a fiction author). High Rocketry Expectations declared to enter the prize after learning from other competitors and figuring out improvements for their designs.

Overall an estimate of $20 million and about 100 000 person-hours were invested in R&D in four years of competition by all teams (Courtland, 2009). Although all the teams performed R&D activities in this competition, the most effective efforts came from Armadillo Aerospace and Masten Space Systems. Masten Space Systems spent about $2.5 million to win the $1 million cash purse (Morring, 2009), and Armadillo Aerospace had already spent at least $3.5 million in its whole development program by 2008 (Armadillo Aerospace, 2008). The rest of the teams were self-funded, lower-budget teams. Within this group of five teams, the data indicate that only Armadillo Aerospace was successful in raising money from corporate sponsors (in addition to the own money invested by its wealthy leader).

The teams sought to develop and manufacture most of their technologies in-house. There is also some evidence of subcontracting (Masten Space Systems) and more significant use of off-the-shelf technologies (BonNova). The organization of R&D activities was also diverse. Armadillo Aerospace was an entrepreneurial team that contributed knowledge and R&D approaches from the software industry. The team referred to its projects with terms such as fast-prototyping and learning-by-testing processes. Masten Space Systems used software development-like iterations for modeling, analysis and testing, what the team calls 'incremental test production'. There were also teams such as Unreasonable Rocket with very simple R&D, 'craft' production organization that drew mostly on learning-by-doing. Another interesting feature of the R&D activities in this prize was the degree of openness of the teams with each other and with the public (Davidian, 2010). The teams used online tools extensively to share their experiences and advances and, most interestingly, may have relied on each other to analyze technical problems and suggest solutions (Pomerantz, 2010b). Team Unreasonable Rocket for example shared its experimental flight permit application with the team TrueZer0, which also made its own permit application publicly available (along with many technology details in the same document) (TrueZer0, 2008). In the understanding of the R&D activities in this prize we shall consider the short development time frames posed by the competition, which in some cases represented a constraint from the technical and bureaucratic point of view. In 2006 for example the teams had only 168 days to build their vehicles between the prize announcement and the competition day (Pomerantz, 2006). In similarly short periods the teams had to obtain experimental permits from the FAA, which represented a significant constraint for all teams.

5.2.6 Technology Outputs

The 12 participating teams developed and tested some technology in these four years of competition. Most of this technology development occurred at medium-to-high TRL (technology readiness levels) (Davidian, 2010). Seven teams participated more than once in the NGLLC. Five teams participated at least three times (including the two winners) and introduced incremental innovations that improved the precision of their flight tests. Throughout the four years of competition, the technical solutions contributed by the teams may have converged to similar designs possibly as the result of mutual learning and imitation (Pomerantz, 2010a). However, program managers suggest that there was no significant duplication of R&D efforts as teams tested diverse solutions to specific problems (Davidian, 2010).

Figure 5.2   Team Armadillo Aerospace and its MOD vehicle at the NGLLC 2007

Assessments by the prize sponsors consider that the NGLLC induced several innovations (Pomerantz, 2007; NASA, 2009c). Armadillo Aerospace for example set new reusability records, introduced an R&D approach based on fast pace modular development, and developed sophisticated computer controls for its vehicles (Figure 5.2 shows a picture of the MOD vehicle built by Armadillo Aerospace to enter the NGLLC in

2007). Masten Space Systems set new standards of efficiency and speed. BonNova developed new patent-pending rocket engine components and set new standards of size and weight. Unreasonable Rocket also developed functional yet not fully performing technology. Some of these technology outputs have not reached commercialization phases and others required further development to be readily available for commercialization. A few teams had a commercial orientation that was incorporated into their designs, using modularity as a design criterion for example, and resulted in patent applications. Other teams considered commercialization possible yet not a priority. The data do not show commercialization of new technologies during the competition. After the NGLLC ended, Masten Space Systems obtained a USA Small Business Innovation Research contract to use its vehicles as a network communications test bed for the Department of Defense (Masten Space Systems, 2009). More recently both Armadillo Aerospace and Masten Space Systems were awarded $475000 to perform test flights of their experimental vehicles under NASA's Commercial Reusable Suborbital Research Program (NASA, 2010e).

5.3   DISCUSSION

These two pilot case studies not only helped in the course of this investigation to test the explanatory power of the innovation model introduced in previous chapters but also contribute some insights to answer the research questions that guide this investigation. To summarize the analysis, the preliminary findings of these pilot case studies can be displayed using a template of the model (Figure 5.3 and Figure 5.4). The evidence shows for example that the motivation of teams is not necessarily associated with a monetary cost-benefit analysis of participation and is more likely related with entrant-level attributes such as perceptions and goals. While the prize incentives of both AXP and NGLLC successfully attracted unconventional entrants, the potential market value of the prize technologies associated to these prizes did not induce the participation of traditional industry players in the space sector. There are at least two possible explanations for this: either (a) the technology incentives have not been sufficiently significant or clearly signaled by the competitions and, therefore, no traditional industry players were interested in these challenges; or (b) other factors such as the risk of participation or the structure of the technology sector may have played a role. In relation to the first explanation the fact that technological targets were defined for a relatively early point along the commercialization timeline may have reduced the attractiveness of both competitions for conventional entrants. In other words there may have

	Motivations	R&D activities	Technology outputs
Ansari X Prize	• **Prize challenge**: be the first non-government organization to launch a reusable manned spacecraft into suborbital space twice within two weeks. • **Monetary reward**: $10 million. • **Prize benefits**: global exposure due to extensive media coverage. • **Prize rules**: open-ended and conducive set of technical specifications; targeted space tourism; focus on requirement of private funding.	• **Perceived incentives**: very diverse motivations; prize money is secondary. The competition offered significant exposure and publicity to teams. Sizable market opportunities were perceived yet directly targeted by only a few teams. • **Risk**: prize participation did not imply additional risk for teams. In general, teams interested in prize money only entered the competition when sponsor guaranteed prize money.	• **Design criteria**: simplicity and low cost were main criteria; re-usability was required by prize rules; reliability and safety were criteria as well. • **Design sources**: teams drew upon existing technologies and new conceptual designs. • **Technology sources**: in-house manufacturing, subcontracting, and off-the-shelf procurement. • **R&D effort**: more than $100 million invested in R&D, external funding attracted. Volunteer effort. • **R&D organization**: diverse R&D approaches; only a more dynamic and entrepreneurial approach succeeded; only a few teams reached stage of development. • **Constraints**: Lack of funding.	• **Technology outputs**: winning entry was a hybrid spacecraft design with innovative configuration, use of new materials. Other innovative concepts were introduced by all teams. Scaled tests introduced new features and materials as well. Hardware outputs concentrated in a few most active teams. • **Commercialization**: potential commercialization was a key driver in design and development and was implicit in prize rules. • **Effect on innovation**: accelerated ongoing R&D activity, attracted new innovative players, and induced conceptual and operational innovations as well.
Characteristics of teams	• **Type of teams**: mostly unconventional teams, incl. pre-existing companies that redirected activities or new startups. Multiple types of partnerships emerged to raise funding. Significant volunteer effort in some cases.	• **Experience/Skills**: all teams attracted individuals with aerospace experience which influenced designs. Overall engineering experience was present in all cases. No team had experience in past competitions or suborbital space flight.	• **Strategy/goals**: commercialization of technology and opportunity to enter the new space tourism market was the initial goal of a few teams. Some independent teams sought to accomplish personal and other organizational goals.
Suborbital manned space flight sector	• **Technology scenario**: technologies likely to have demand if they are affordable and safe. First space tourism experience not seen until 2001, after prize announcement.	• **Technology-related incentives**: sizable market opportunities in space tourism market through suborbital flights. • **Technology gap**: though technology already existed, significant effort was needed to make spaceflights more affordable and safe.	• **Sector structure**: sector that emerged along with the competition; comprises new players, mostly small companies that re-directed activities or were just created. Regulatory framework required modifications to allow flights.

Source: Author's analysis based on secondary data described in the text and model introduced in previous chapters.

Figure 5.3   Summary of analysis of the AXP

	Motivations	R&D activities	Technology outputs
NGLLC	• **Perceived incentives**: prize money, potential reputation (which may lead to commercialization), and self-fulfillment are drivers of engagement and R&D effort. • **Risk**: teams internalize risk of participation and balance with risk of technology development since prizes are a mean to accomplish their goals.	• **Design criteria**: simplicity and low cost were main criteria; diverse set of approaches led to unconventional criteria as well (e.g. modularity). • **Design sources**: teams contributed non-space knowledge and approaches to R&D. • **Technology sources**: in-house manufacturing predominated. • **R&D effort**: teams invested $20 million and 100 000 person-hours in pursuit of the prize purse. • **R&D organization**: very small, flexible teams with informal yet novel approaches to organize R&D and learn during the process. • **Constraints**: short development time frames.	• **Technology outputs**: new components and subsystems, new standards of reusability, speed of development, and efficiency. • **Commercialization**: in some cases, technology outputs had commercial orientation since design stage. • **Effect on innovation**: triggered new R&D activity in a narrowly defined technology sector with no active developments.
Characteristics of teams	• **Type of teams**: mostly unconventional teams created as independent, volunteer R&D teams to join the competition. A number of them participated in several instances of the competition.	• **Experience/Skills**: in general, diverse engineering experience, aerospace experience only in a handful of cases. Importance of learning-by-doing during the competition. Participation in past competitions had influence.	• **Strategy/goals**: pursuit of a challenging goal related to personal interests in most of the teams. A few teams used the competition to promote activities and, eventually, commercialize technologies.
Space exploration research and procurement sector, VTOL vehicles	• **Technology scenario**: opportunities to produce and comm. affordable and productive Moon exploration and suborbital flight technology in the medium- or long-term. • **Technology gap**: existing yet not efficient/affordable technology; development of more precise, effective, and affordable vehicles.	• **Technology-related incentives**: Moon exploration is a target of space agencies; subcontracting by them (e.g. NASA) or large corporations (e.g. Northrop Grumman) are a direct commercial opportunity; other scientific and corporate applications exist for related services and technology.	• **Sector structure**: space agencies, few large corporate players, and large contracts have traditionally dominated this sector (e.g. Northrop Grumman, which has been linked to lunar programs since the 1960s).

Additional NGLLC details:
• **Prize challenge**: build and fly VTOL vehicle that simulates the flight of a vehicle on the Moon.
• **Monetary reward**: $2 million divided into two levels and 1st and 2nd place.
• **Prize benefits**: exposure to lots of general public and NASA and corporate officials at competition site.
• **Prize rules**: narrowly defined technology target; detailed technical specifications, government funding not allowed.

Source: Author's analysis based on data described in the text and model introduced in previous chapters.

Figure 5.4   Summary of analysis of the NGLLC

been a perception of no immediate, concrete markets for technologies that would not yet be flight-ready by the end of the prize. In relation to the second explanation there may be other sources of risk not related with technology development such as the potential ruin of business reputation if an established player loses to an informally organized R&D team. This suggests that unconventional entrants might be less risk averse or more optimistic in their forecasts than established industry players. Moreover, the participation of traditional entrants may also depend on the structure of the industry sector. This explanation fits particularly well in the AXP, which was linked to an emerging market with no established, specialized companies. In the NGLLC case there existed large corporations or specialized firms, yet only 12 mostly unconventional teams entered in four years of competition. This suggests sector entry barriers such as lack of knowledge, skills or funding. More generally these explanations concur with Christensen's disruptive innovation model (Bower and Christensen, 1995) in which new market entrants challenge established industry incumbents by offering simpler alternatives that are good enough for underserved niches. From this perspective, prizes may set conditions for new R&D performers to enter mature markets if they succeed in finding better technological solutions that target markets related with the prize technologies.

Both competitions involved diverse R&D approaches and teams faced a number of constraints. The investigation of the effect of time and budget constraints requires more in-depth investigation and richer data. The effect of each constraint should be explored in more detail to discover prize specific relationships. Simpler and low cost approaches for example are likely to relate to the lack of up-front funding. The use of existing technologies is likely to be the response to shorter development lead times. In-house manufacturing may be the response to the need to control technology development and prevent unexpected events. More clearly unorthodox R&D approaches result from engaging entrants with diverse backgrounds. Though the aerospace industry has been characterized as multidisciplinary since its emergence (Bromberg, 2000) these competitions allowed a full deployment of capabilities and approaches not generally seen in instances of traditional technology procurement such as those that involve NASA and prime contractors. The NGLLC offers a concrete example of this influx of knowledge from other sectors. Team Armadillo Aerospace, one of the winners, was observed to bring 'the dynamism of software development to aerospace projects' as its members had mostly IT industry background (SpaceRef.com, 2008). The rules of the AXP were also characterized as open-ended and conducive to develop the prize technologies (Linehan, 2008) and may have enabled the implementation of creative, non-space industry ideas.

These prizes leveraged significant R&D investment and induced innovations that would have not occurred anyway, at least by the prize deadlines. The caveat is that the competitions were linked to significant technology incentives and, most importantly, ongoing R&D processes. Experts and scholars also consider that these prizes 'provided a focus for efforts that people already had under way' (Greason, 2010) or simply motivated the last effort to come up with the final solution (Saar, 2006). While a significant breakthrough was observed in the AXP, the technologies used in the winning entry may have been already under development. Still the prize set a concrete technological goal and showcased the accomplishments of Scaled Composites and a handful of other unconventional teams (e.g. DaVinci Project, ARCA). These runner-ups also contributed R&D effort that would have not existed at that point without this prize (since those teams were created to enter the competition or the competition was among their most important goals). In the case of the NGLLC, Masten Space Systems and Armadillo Aerospace were pursuing similar R&D before the prize announcement, but their innovations specifically addressed the prize target. The rest of the teams were mostly new-to-industry, volunteer teams performing new R&D activities.

In sum the immediate contribution of these competitions has been the definition of the characteristics of the innovations to be achieved and hence the focus of R&D efforts. Their effect on innovation can be associated with the factors anticipated in Chapter 3. More significant prize incentives induce more significant R&D efforts, wider technology gaps are more likely to induce the development of novel technologies and more open-ended challenge definitions enable innovative approaches and designs. In practice however the effect of these factors is complex and interrelates with the characteristics of the entrants, the technological field and the broader context to produce prize outputs in a continuum that goes from non-innovative production to technological breakthroughs, passing through incremental and other less significant innovations. In the longer term the competitions may have other outcomes as well. Some teams and new start-ups for example continued with their R&D programs to develop the space tourism and scientific payload delivery segments, and these activities attracted new follow-up investments (Belfiore, 2007). These competitions also demonstrated the capabilities of the private space industry and raised public interest in space activities.

The pursuit of commercial opportunities is apparent in both competitions. In general the innovations described earlier were implemented for own performance improvement in progress toward the prize target, and commercialization was only observed after the prize deadlines, yet the ability of teams to retain the IP on their technologies has been a key factor.

Discretion on the use of IP facilitates fundraising and future commercialization to repay investments or create a sustainable business. Contracts signed by entrants of both competitions demonstrate how significant the potential technology incentives linked to prizes may be and suggest that prize winners are the most likely to realize those commercial opportunities to full extent. Patent-pending developments by other teams also suggest other potential future commercialization activities.

NOTES

1. The classification of prize entrants in the AXP and NGLLC is based on their industry experience (i.e. space agency/industry experience) as in the GLXP case. However, a lack of primary data on the experience of individual team members only allowed a classification of entrants at the organizational level. In other words this required the author to determine whether there was a relationship between pre-prize activities of the team as a group or organization and the prize challenge.
2. Other examples of conventional entrants (not included in this investigation due to the lack of data) are Interorbital Systems, Kelly Space and Technology, and Pioneer Rocketplane Inc.
3. The company has also sought to support NASA on the development of Altair, a lunar lander that is expected to place four astronauts on the Moon by 2020 (XPF, 2009; Northrop Grumman, 2007).
4. Other entrants such as Paragon Labs, Acuity Technologies, Speed Up, and TrueZer0 (not included in this investigation) were small companies that re-directed their activities.

6. A closer look: the Google Lunar X Prize

6.1 THE GLXP PRIZE

On 13 September 2007 the XPF launched the GLXP with the sponsorship of Google Inc. The competition offers up to $30 million in prizes to the first entrant that lands a robot on the surface of the Moon, among other secondary technological targets, by 31 December 2015. The XPF designed this prize to 'accelerate technology developments supporting the commercial creation of multiple systems capable of reaching the lunar surface and performing operations over an extended period of time.' Other secondary goals include educating about the benefits of space and Moon exploration; raising interest in science, technology, math and engineering; supporting a new generation of engineers and entrepreneurs; and opening the space frontier to new ideas and participants by lowering costs. This competition is also expected to have a broader sociological impact since half of today's world population was not alive at the moment the last NASA Apollo mission visited the Moon. We have to consider that the GLXP, as well as other competitions launched by the XPF, is also designed to have multiple back-end business markets that can be supported by the technologies developed in pursuit of the prize. In this regard the XPF's criteria for prize design include that entrants should ideally be able to continue working on the commercial development of the prize technologies after the end of the competition and also differentiate their strategies through different business models. The advancements made by the GLXP entrants for example may eventually allow NASA and other space agencies to save money and expand the capabilities of future robotic and human missions to the Moon (Pomerantz, 2010a).

The announcement of the competition came at the annual *Wired* magazine's NextFest event of 2007, in Los Angeles, California. The XPF considered that the event would make the GLXP visible to the 'right kind of people' or 'nontraditional competitors' which include 'dot-com billionaires, open source people and a million different other kind of people who might see this as interesting enough to get involved' (Pomerantz, 2010a). The usual contact that the XPF maintains with established industry players in conferences or for the organization of other prizes helped to reach out the broader aerospace community. The almost simultaneous launch of Kaguya, a Japanese lunar orbiter, by the Japan Aerospace Exploration Agency, might have brought further attention to space activities in general and the GLXP in particular. Since the announcement of the prize the XPF maintains the GLXP official website with information on the competition, its rules and its contenders and updates on prize developments (the XPF requires teams to use the website to publish updates on their general activities).[1]

6.1.1 The Challenge Definition, Prizes and Rules

The GLXP challenge basically requires teams to launch a spacecraft from Earth to Moon, land it on the Moon, deploy a rover (or equivalent unit) to traverse 500 meters, and collect and send high-definition (HD) video footage back to the Earth before prize expiration. The cash purse that the prize sponsors will award for that accomplishment is divided into a $20 million Grand Prize,[2] a $5 million Second Place Prize and other prizes totaling $5 million. The Grand Prize will be awarded to the first team to complete all of the mission requirements as described in the following paragraphs. The Second Place Prize will be awarded to the second team to complete those requirements. At sole discretion of the XPF, this prize may also be awarded (as some kind of consolation prize) to a team that accomplishes most of the requirements to win the Grand Prize but, due to unforeseen reasons such as mechanical difficulties, ultimately fails to meet all the mission requirements (in which case the Grand Prize would be still available). The Second Place Prize may not be won by the Grand Prize winner. The Bonus Prizes will be awarded to the team or teams that successfully complete the bonus requirements and the Grand Prize or Second Place Prize mission requirements.

To be able to claim the Grand Prize, entrants have to meet these requirements:

1. *Landing*   The team must land its craft on the surface of the Moon.
2. *Mobility*   The craft or a single secondary vehicle carried by the craft must move a distance of at least 500 meters along the surface of the Moon in a deliberate manner (on, above, or below the lunar surface, yet with straight line displacement capability for that distance).
3. *Mooncast transmission*   The craft or a secondary vehicle must transmit from the surface of the Moon two eight-minute high-definition (HD) videos (an Arrival Mooncast and a Mission Complete Mooncast) exclusively for the XPF or its partners.

4. *Data uplink*   The team must transmit to the craft or secondary
    vehicle while on the surface of the Moon as much as 100 kilobytes of
    data provided by the XPF for later transmission back to Earth.
5. *Payload*   The craft or secondary vehicle must carry an XPF payload
    of about one per cent of the craft or secondary vehicle's dry mass, with
    a minimum mass of 100 grams and maximum mass of 500 grams.

There are also a number of bonus and diversity prizes totaling $5 million.
The $4 million Apollo Heritage Bonus Prize is for the first team that takes
imagery and video of an Apollo site and of a historical artifact associated with the Apollo mission. The $1 million Heritage Bonus Prize is for
the first team that takes imagery and video of a historical site of interest
including footage of an artifact associated with a previous mission to the
Moon other than the Apollo missions. The $2 million Range Bonus Prize
is for the first team that moves its craft or a secondary vehicle along the
surface of the Moon for no less than 5 km. The $2 million Survival Bonus
Prize is for the first team that successfully operates its craft or a secondary
vehicle on two separate lunar days. And the $4 million Water Detection
Bonus Prize is for the first team that provides scientifically conclusive
proof of the presence of water on the Moon.[3] The GLXP also offers a
$1 million Diversity Award to the team that, in the opinion of a panel of
experts, has made the greatest attempts to promote diversity in the fields
of science, technology, engineering and mathematics.

   The rules of the GLXP are established by the Prize Guidelines and the
Master Team Agreement (MTA), a binding contract that every team has
to sign to enter the prize. While the former focus on technical requirements and are more informally written to be understood particularly by
engineers and entrepreneurs, the latter sets guidelines and legal provisions
that regulate a number of key aspects of the competition. Two key rules
are associated with IP and R&D funding. The first is a provision that
allows teams to retain the IP rights associated with the design, manufacture and operation of the spacecraft, secondary vehicles and subsystems.
This is essential to enable technology commercialization and fundraising
as discussed later on. On the other hand there is a limit on the use of public
funding and other resources. According to the rules, funding from government sources cannot constitute more than 10 per cent of the total funds
available to a team to compete, including support received as cash and
the cash-equivalent of in-kind support. In addition, teams are not allowed
to purchase preexisting hardware from sources such as museums, space
agencies or defunct companies, unless equivalent or superior replace-
ment products are commercially available. Teams are still allowed to use
governmental facilities, personnel, hardware or information previously

developed by a government organization, if access is available to other teams as well. Government personnel are allowed to work for a team so long as they are working outside of the scope of their government employment.[4]

To enter the GLXP the teams had to submit an application package between September 2007 and December 2010 with diverse information about the team and its members, finances and mission plan. From the time the XPF opened the registration through the end of the calendar year 2008, there was a $10 000 registration fee. For calendar years 2009 and 2010, the fee was raised to $30 000 and $50 000, respectively. Teams could obtain a temporary extension of the $10 000 and $30 000 rates by filing a Letter of Intent to Compete, but no such extensions were possible at the end of 2010.

6.1.2 The Idea of a Lunar Exploration Prize

The original idea of a competition for lunar exploration emerged from NASA. In 2006 NASA requested studies from Paragon Space Development Corporation and the XPF on the impact of an incentive prize for robotic lunar exploration. Paragon's study was a parametric analysis of the cost of the cheapest possible lunar surface mission, which would suggest how big a prize purse would need to be to incentivize someone to actually go out and pursue that mission. XPF's study approached the topic from an opposite angle. The purpose was to find out what would be the most the agency could get out of the part of the community of developers that might pursue the mission. The results of these studies suggested a prize purse of about $20 million for that kind of challenge, considering only a minimum of USA teams and the requirements to land on the lunar surface and take pictures, but not move (Pomerantz, 2010a). The Constellation program for returning humans to the Moon was a priority for NASA at that time, and a prize like the GLXP would contribute significant data for that program. But the authorization for a cash prize of that amount would have been difficult to obtain for NASA and the competition would have been restricted to USA teams only. Further conversations between the XPF and experts were then followed by Google's interest in supporting the competition. By this time the idea of requiring some kind of surface mobility was considered. Mobility was expected to be the enabling technology that would make a big difference on the type of mission. Covering a distance of 500 meters on the Moon surface was considered a demonstration of control and the ability to move in any direction. The XPF then conducted an industry survey of executive-level individuals from industry, academia and government to assess the feasibility of a GLXP-like mission.

Based on a hypothetical $20 million cash purse, 70 per cent of the respondents considered that a mission including a lunar rover and video of an Apollo landing site was feasible without support from NASA. The mission achievement lead time from prize announcement was estimated at about four years in that survey (Pomerantz, 2006).

To the author's knowledge, there is no other competition offering a similar amount of cash purse to accomplish a robotic exploration mission. A prize that relates only to some extent is NASA's National Lunar Robotics Competition. In October 2009 three teams claimed a total of $750000 in prizes at this competition. Competitors were required to use mobile, robotic digging machines capable of excavating at least 330 pounds of simulated moon dirt, known as regolith, and depositing it into a container in 30 minutes or less. The rules required the remotely controlled vehicles to carry their own power sources and weigh no more than 176 pounds.[5]

6.1.3   Organizer, Sponsors and Partners

The XPF is the organizer of the GLXP. This foundation is an educational, non-profit corporation established in 1994 to inspire private, entrepreneurial advancements in space travel. This is the largest single international prize ever offered by the entity but not the only one. The XPF also organized the AXP and NGLLC and has several other prizes in concept development process, or already launched in areas such Energy and Environment, Life Sciences and Education and Global Development (XPF, 2011b). Google, Inc., the sponsor of this prize, has been a long-time supporter of the XPF and the approach of using competition to stimulate the private sector to achieve important goals more quickly and affordably than previously possible. The GLXP in particular is not related to Google's core business, yet the company expects to contribute to start what they call 'Moon 2.0', a new era of lunar exploration that will be more participatory and more sustainable than the first Moon race that ultimately led to the Apollo missions (XPF, 2008a).

The XPF has also partnered with other organizations that sponsor this competition and offer products and services to participant teams. They are considered preferred partners and support teams in different technology areas of the GLXP mission. Space Exploration Technologies, for example, offers a discount of 10 per cent of the launch costs for all launches on its Falcon 1, Falcon 1e and Falcon 9 rocket vehicles. The SETI Institute offers free use of its Allen Telescope Array for receiving data from the surface of the Moon for the first seven Earth days of operations on the lunar surface. The Universal Space Network offers a 50 per cent discount on communication services for the spacecraft while in transit to the Moon and for 30 Earth days of operations on the lunar surface. Space Florida, an Independent Special District of the State of Florida charged by the Florida Legislature with promoting and developing Florida's aerospace industry, offers a Bonus Prize of $2 million to the winner of the competition that launches its spacecraft from Florida. And Analytical Graphics Inc. provides one license of its STK package for complex mission planning from launch to landing, valued at $150 000 each, to all registered GLXP teams free of charge.

The GLXP is not officially supported by any government or space agency. NASA has not officially endorsed or supported the GLXP, yet congratulated the initiative of the XPF particularly with regard to the efforts of the Foundation to engage youth and inspire students to pursue careers in science, engineering and other fields related to space exploration (Griffin, 2007). JAXA, the Japanese Aerospace Exploration Agency, also congratulated the XPF for this initiative (Tachikawa, 2007).

6.2 THE CONTEXT OF THE COMPETITION

6.2.1 The Space Sector Structure

Government-led efforts have historically driven space activities. The USA, the Soviet Union/Russia, Japan and some European countries have traditionally been the leaders, but other countries such as China and India have lately expanded their space activities and gained a share of the sector. In the USA agencies such as NASA and the Department of Defense have been the main drivers for civil and military space development, respectively. The USA government is still the largest single customer for technologies and services such as launch vehicles. Large corporations such as Northrop Grumman, McDonnell Douglas and Boeing have been involved in space developments since the early days of spaceflight, as prime contractors for government agencies. Other companies and universities have also been involved through grants, contracts and cooperative agreements. This mostly publicly funded space sector has evolved considerably since its origin in the 1950s and 1960s, particularly in terms of budget and organizational complexity. Government space agencies have generally grown in size and become centralized, bureaucratic and less productive organizations with substantial fixed costs, and the private space industry has consolidated into a few large players.

The organization of R&D in this industry has been influenced significantly by subcontracting and supervision policies of government agencies. Both agencies and large contractors have built strong in-house capabilities, extensive control systems to manage large and multiple projects, and complex hierarchical structures with division of labor between multiple R&D centers with multi-disciplinary and multi-team structures (McCurdy, 1994; Bromberg, 2000; Cucit et al., 2004; Petroni et al., 2009). Large companies involve suppliers and other partners to source specialized competencies, technologies and knowledge, but maintain the design authority, administer the development effort, and assemble, test and market the products (Baird et al., 2000; O'Sullivan, 2003). The increasing complexity of systems leads to both increasing programmatic risks and high task interdependency between development teams.[6] As a result project management and communications have become critical factors to project success and required more complex and bureaucratic organizational forms. Accidents in space projects have also led agencies and companies to focus on risk management procedures, which increase bureaucracy and control even further to prevent failures (Kranz, 2000).

Space projects are typically large scale and involve large spacecraft and payloads. They are generally one-of-a-kind projects that integrate diverse technologies and have long development cycles that can exceed 10 years. Their complexity and cost have increased to the point that no one organization or company can afford to tackle space exploration projects alone (Bugos and Boyd, 2008). Space flight and exploration systems have also increased their performance and reliability over time. Space technologies are advanced either internally from research to implementation in own government missions or by contractors that enter in partnerships or receive support to advance technologies with diverse levels of maturity. Spacecraft are generally based on more expensive heritage parts that draw on designs and processes that have been successfully used in other projects, including the conventional rockets used to launch spacecraft and payloads. Project cost is a dependent variable in this context and a trade-off between cost and schedule is dictated by excessive overhead costs, which results in very expensive technologies when there are long development cycles. Sometimes, when systems are finally built for long-term missions, other cheaper commercial solutions become readily available (CBO, 2004).[7] Delays and cost over-runs have also been common in space programs. NASA's past robotic scientific missions for example had cost and schedule growth rates of 20 per cent or more due to the inadequate definition of technical and management aspects, program funding instability, program re-designs, technical complexity and budget constraints. The development cycles for robotic exploratory missions (60 months or shorter) have generally been shorter than typical Earth-orbiting missions due to more constrained launch windows, but have historically failed twice as often as other space programs (Bitten, 2008).

There have been significant changes in the space sector since the 1990s. Most importantly, new entrepreneurial companies have started to enter this industry to offer more affordable solutions for commercial space development in market segments such as space tourism and payload delivery (Gump, 1990; Cucit et al., 2004). This has occurred in part due to space policies and regulations with more commercial and entrepreneurial approaches which contributed to the emergence of a new space sector and forms of R&D organization (Culver et al., 2007). In the USA NASA's programs have also contributed significantly to develop an emerging commercial space sector by supporting some of those new companies.[8] Industry experts suggest three other factors that explain the space entrepreneurship phenomenon:

1. *Overall sector evolution*
   The constant need to advance technology in diverse areas linked to aerospace has pulled an increasing number of new small companies to provide technologies such as software, energy and power systems, which have become key subsystems in aerospace.
2. *Feasibility demonstration projects*
   Projects such as the 1990s Delta Clipper DC-X[9] led many people to revise what the conventional wisdom said about what is and is not possible in aerospace development. New ways of doing business, managing companies and structuring programs have been brought into the aerospace industry since then, from outside.
3. *Generational factor*
   The executives of new space companies tend to cluster in a fairly small range of ages, as they were either very young or not yet born when NASA's Apollo program finished. These executives may have had a common sense that the potential and promise shown in the last days of the NASA space race were unfulfilled and potentially unrealized. Therefore, despite having non-aerospace careers, these executives have never lost interest in space and at some point realized that they can contribute to fulfill such promise and potential.[10]

Government agencies, large contractors and new actors in the space sector have also implemented new mission approaches known as small missions. These use combinations of the latest technology with commercial off-the-shelf (COTS) technologies (or other mature technologies) that enable increasing capabilities in smaller spacecraft due to their general miniaturization. The lower cost of COTS technologies has allowed engineers to compensate for their lack of space heritage by engaging in much more rigorous testing at a much earlier stage in program development. Small missions are also generally associated with aggressive and early prototyping and testing, rapid development schedules and focused objectives. All these factors have led to faster turnarounds (development cycles of 12–36 months) and more missions with significantly lower budgets and programmatic risks, although still with mass and size constraints. This kind of mission has also allowed more, and more diverse, actors to enter the sector, including small companies, universities, countries not traditionally involved with aerospace development and even amateur or other non-profit initiatives (Marshall et al., 2007; Bonin, 2009).[11] Two illustrative examples help to better understand the small mission approach. The first is STRaND-1, a small satellite containing a smartphone payload that will be launched into Earth orbit in 2012. This is a low-cost satellite that is being built in engineers' free time by the University of Surrey and Surrey Satellite Technology Limited (SSTL) to demonstrate the advanced capabilities of a satellite built quickly using advanced COTS components (SSTL, 2011). The second is Copenhagen Suborbitals, an open source, non-profit initiative with the goal of launching a human being into space. This is an amateur effort but also involves individuals with space/industry experience. Based entirely on sponsors, private donors and part time volunteer efforts (about 20 people in 2011), this organization has performed dozens of engine tests and accomplished its first test flight in mid-2011 with an annual budget of less than $100 000 (Copenhagen Suborbitals, 2011).

Sector regulations and other institutional settings certainly affect the development of space activities and how this industry is organized. Most importantly, USA citizens and organizations that work with certain aerospace and defense technologies are required to abide by the US International Traffic in Arms Regulations (ITAR) which prohibits companies to export technologies that are considered inherently military in nature or have dual-use. Among other effects, this type of regulation has generally limited international cooperation and forced some foreign agencies and companies to introduce ITAR-free designs in their spacecraft (Hudson, 2008). There are also other factors related with intellectual property protection that may eventually affect companies that successfully launch their missions and seek to commercialize services. Some maintain for example that 'common terrestrial legal practices' such as licensing terms may not be suitable for outer space operations (Hudgins, 2002; Kleiman, 2010).

6.2.2   Technology Gaps

To analyze the technology gaps that GLXP entrants have to close we shall consider the prize challenge in relation with the state of the art of space technologies linked to planetary robotic exploration.[12] The activities in this field specifically aimed at exploring the Moon started about five decades ago. In 1966 the Luna 9 unmanned spacecraft, part of the Soviet Union's Luna program, became the first spacecraft to achieve a soft landing on the Moon (or indeed any planetary body other than Earth) and send photographic data back to Earth. Agencies from the Soviet Union, USA, European and Japanese governments have accomplished near 50 successful missions to the lunar surface or its orbit since then (Schrunk et al., 2008). The last successful USA lunar lander was Surveyor 7, the fifth and final spacecraft of the Surveyor series sent to perform a lunar soft landing in 1968. The last spacecraft to land on the Moon was the Soviet Luna 24 in 1976.

The GLXP teams have to consider not only the requirements of the prize to be able to claim the prize, but also other more general technical challenges involved in any kind of robotic exploration mission. The GLXP challenge definition is very open-ended and does not specify the technologies or the technological means the teams have to use, but it does involve a set of minimum technological capabilities defined in previous sections as soft landing, surface mobility, Moon video broadcasting, data uplink and payload.

The first technical challenge that all space missions to the Moon face is the Earth-to-Moon launch. Teams first have to reach the Earth's orbit and then transfer their spacecraft to the Moon. Teams may either build their own launch rocket or buy already proven commercially available solutions.[13] Both options are equally expensive, but the development of a rocket from scratch demands time that is not available in this competitive context. There are readily available commercial launch services to reach the Earth's orbit, such as those offered by SpaceX with its Falcon 1e and Falcon 9 rockets, but they are very expensive ($10 million and $50 million, respectively; the bigger the spacecraft mass, the more expensive is the launch rocket) (SpaceX, 2011a, 2011b).[14] The payload requirement of the prize (to carry an XPF payload of about one per cent of the craft or secondary vehicle's dry mass) is not in principle a significant problem for most of the teams as they plan to deliver bigger payloads.

Soft landing on the Moon is the next big challenge and possibly the most challenging part of this mission from the technical viewpoint. Since the Moon has no atmosphere, common atmospheric descent methods (e.g. supersonic parachutes) are not suitable, and completely propulsive methods are needed. Cheap and lightweight systems for landing on the Moon (e.g. airbags) have fairly poor accuracy. More accurate landing is possible with more expensive and heavier systems, but the teams have to consider that the bigger the mass of the spacecraft and landing system, the bigger and more expensive the launch service. Soft landing on the Moon was already achieved by past robotic missions and even by NASA's manned Apollo missions, which is a very important precedent and source of knowledge for teams. Systems that would significantly help to increase landing accuracy include for example autonomous landing and hazard avoidance technologies (such as 3D Imaging and Flash LIDAR technologies) are at low-medium TRL, and may not be readily available to teams due to their cost and maturity. Other sophisticated landing technologies such as warm gas descent engines still have medium levels of maturity. Interestingly this is the only part of the mission that cannot be tested prior to launch because, for example, on Earth it is not possible to neutralize the gravity and recreate the vacuum-effect that affects the performance of thrusters and engines used for the lander's descent.

Mobility on the Moon surface is the next challenge after lunar landing. The technical complexities associated with this emerge mainly from extreme temperatures and the presence of lunar dust on the Moon surface. Huge temperature fluctuations (from 123 Celsius during the day to minus 233 Celsius at night) can affect and damage hardware. Teams have to figure out methods to maintain appropriate operational temperatures for sensitive equipment (such as electronic components) and develop mechanisms and materials that support abrupt temperature changes. Mechanisms such as motors and robotic arms and components such as bearings, seals and lubricants that work under dust and extreme temperature conditions are in early stages of development and may not be readily available. To cover a 500 meter distance without requiring complex equipment for lunar night hibernation, the GLXP lunar vehicles should be able to traverse such a distance in less than a Lunar day (i.e. less than 14 Earth-days). Past robotic missions[15] and mobility technologies that have been available since the 1970s with different levels of maturity indicate that this is a plausible goal. Teams can draw for example on some mechanically complex but mature technology designs such as wheeled robot mechanisms that are spin-offs of terrestrial applications and were used in missions to the surface of the Moon and Mars. Other less conventional designs such as hoppers and legged systems are purely for space application and generally are in low-to-medium TRLs.

Another challenging technical issue that arises after a successful Moon landing is how to power the spacecraft. Solar powered Moon vehicles for example can take advantage of Lunar days that last 14 Earth-days, but increasing power needs also increase solar panels' mass. Lunar nights that also last 14 Earth-days require alternative means to power systems. More advanced power sources such as primary fuel cells and regenerative fuel cells are under development or in early-stage design and may not be available to teams. In any case backup batteries are required to ensure power under varying operational conditions. The teams may use batteries that are commercially available for other terrestrial applications, but which would require adaptation and testing efforts to guarantee a satisfactory operation in the mission.

Teams are also required to broadcast HD video back to Earth. This does not involve complex scientific equipment such as that seen in past robotic missions, but requires special cameras that satisfy the specifications of the prize rules and a data-link for image and video data transmission. The rules require capabilities such as being able to capture full 360° views of the landing site. This may ultimately involve complex technical mechanisms, as cameras may need to include more motors, for example to be able to capture images and video from multiple views. Moreover, while video capabilities are well developed for Earth applications, this would be the first time that HD video is transmitted from the Moon. There are on the other hand some technical challenges in Earth-Moon-Earth communications. The communications bandwidth is limited by virtue of power issues, limited aperture size and access to Earth-bound deep-space networks.[16] This could not only affect the HD video broadcast but also make it more difficult to accomplish the prize data uplink requirement. Moreover, traveling at light speed, information takes 2.5 seconds to travel from Earth to Moon. This kind of delay can affect the remote driving of lunar vehicles for example.

An overview of past planetary missions is instructive to fully understand the technical and project management implications of a GLXP-like challenge. We can take as a reference a number of NASA's Moon and Mars government-led robotic programs (Table 6.1). These programs represent approaches relevant to the GLXP mission except for the Ranger missions that used hard landings and mission goals that generally included both exploration and scientific components. Past missions have relied mostly on the deployment of landers and surface-based, wheeled single units (i.e. rovers) with exploration capabilities constrained in terms of time, distance and operational environment, yet still superior to the GLXP's mobility requirements. The scientific component has generally been related with measuring and analyzing surface and environmental features and included capturing video and images of the visited bodies. The exploration goals have been related with landing in certain areas of interest and inspecting the surface using rovers with Earth-based control to search for targets of scientific interest (e.g. rocks). The budgets of the programs ranged between $170 million and $850 million if the costs to build, launch, land and operate the spacecraft are considered.[17] The Mars Pathfinder, for example, a mission designed primarily to demonstrate a low-cost way of

Table 6.1 Selected robotic planetary exploration programs

Program	Mission	Year[a]	Description	Mission duration[b]	Mass (kg.)[c]	Traverse distance[d]	Program go-ahead to launch period	Development lead times	Budget (current US$)	Scientific component
Ranger (NASA)	Ranger 7–9	1964–65	Hard landers	Impact	n/a	–	1959–65	First test launch in 1961	$170 million (9 spacecraft)	First close-up photos of the lunar surface
Surveyor (NASA)	Surveyor 1–7	1966–68	Soft lunar landers	30–210 days	n/a	–	1961–66	60 months	$469 million (7 spacecraft)	Sensors for surface/ environ. measure-ments
Luna (USSR)	Luna 17	1970	Lander and wheeled rover Lunokhod 1	90 days	900	11 km (speed 1–2 km/hr)	n/a	n/a	n/a	TV cameras and other scientific instruments to explore the surface and return pictures
	Luna 21	1973	Lander and wheeled rover Lunokhod 2	56 days	840	37 km (speed 1–2 km/hr)	n/a	n/a	n/a	
Mars Pathfinder (NASA)	Mars Path-finder	1997	Lander and lightweight wheeled robotic rover named Sojourner	120 days	11	100 m (speed 0.036 km/hr)	Oct 1993– Dec 1996	38 months	Capped $150m project implemen-tation plus $22m rover	n/a

	Program years[a]		Duration[b]	Mass[c]	Distance[d]			Initial cost	Instruments
Mars Exploration Rovers (NASA) Spirit	2003–today	Mars rover to search for and characterize a wide range of rocks and soils	3 Jan 2004–today	185	7.7 km	July 2000–June 2003	32 months; 4 years shorter than historical for this type of development	Initial cost to build, launch, land and operate rovers was $299m (each); grew to $420m; 'open check book' mission	Panoramic camera and several scientific instruments
Opportunity	2003–today	Mars rover to search for and characterize a wide range of rocks and soils	24 Jan 2004–today	185	20+ km	n/a	33 months; 4 years shorter than historical for this type of development		

Notes:
a. Program years
b. Duration of planetary exploration
c. Spacecraft mass (including lander and rover)
d. Distance traversed in Moon/Mars; n/a: not available. Monetary figures in US dollars.

Sources: Spear (1995); NASA (1997); CBO (2004); Zakrajsek et al. (2005); MSNBC.com (2007); NASA (2010g, 2010f).

89

delivering science instruments and a rover to the surface of Mars in 1997, had capped costs of about $200 million. The Mars Exploration Rovers (MER) were developed in exceptionally short periods (less than three years) but at the cost of significant budget increments that in total exceed $420 million (Dornheim, 2003).

Many of the technologies that the GLXP teams need to accomplish the mission are in fact commercially readily available and include both space-rated and less expensive non-space components. Some high performance technologies used in past planetary missions, however, are only available to government agencies. These include subsystems such as the Radioisotope Heater units used to maintain sensitive electronic equipment at normal operation temperature in deep space or other planetary environments.[18] Significant knowledge documented in NASA's Apollo and other planetary exploration programs is also publicly available, from space agencies and other organizations. The NASA's Technical Reports Server (NTRS), for example, is a rich online source of technical documentation.[19] The pursuit of an Apollo-like soft landing or similar type of project under very different cost conditions might however represent a significantly different technical problem in both quantitative and qualitative terms, particularly if the commercial viability of technologies is sought.

6.2.3   Technology Scenarios

The development of technology scenarios for the next decade highlights important changes in robotic exploration mission approaches and significantly superior capabilities in space systems relevant to a GLXP-like mission. A better understanding of these scenarios can help to distinguish prize effects from ongoing industry trends.[20] New mission approaches for example are expected to provide higher scientific returns and increase the reliability of space programs by deploying multiple units per mission. Total mission costs are likely to decrease and the use of multiple units organized in tiers or hierarchies will reduce the costs and risks involved in catastrophic damage of single units. Exploration capabilities will be expanded with the introduction of new legged and hopping mobility technologies and more intelligent coordination and reconnaissance capabilities to access and explore multiple sites per mission. Future missions will also increase their duration to total several months or even years and reach longer surface exploration distances and optimal paths in autonomous mode. Science capabilities are also likely to increase as new on-site assistance for decision-making is added to current-day scientific instruments.

Next-generation robotic exploration technologies on the other hand will be notably more capable than those currently available to accomplish the GLXP challenge. Propulsion systems for Earth-to-Moon transfer will feature higher thrust and payload capacities. Systems lifetime and re-usability are also likely to increase. New technologies will include for example more efficient ion motors, which are currently at medium TRL. New landing technologies will feature more advanced avionics and navigation systems such as 3-D Flash LIDAR imaging systems for safe landing and hazard detection and avoidance (equivalent but less capable technologies were used in the lander that deployed the MER rovers in Mars). Warm gas thrusters (currently at medium TRL) will allow softer and more accurate landing. New mobility technologies will include hazard avoidance systems for quick assessment and visual estimation of the properties of the terrain. New autonomous systems will resolve path choices on their own, outperforming current-day local map-based obstacle avoidance systems. Mobility mechanisms (today predominantly wheeled and vulnerable to extreme environments) will include high performance mechanisms such as motors that work under extreme temperature and dust conditions. New communication systems will increase downlink rates hundredfold to 200 Mbps or higher rates.

The deployment of some of these technologies might occur over the next 10 years with the accomplishment of new government missions. Several countries and their space agencies are planning lunar landings over the coming decade, including the European Space Agency (ESA), the Russian Federal Space Agency (Roscosmos) and the Indian Space Research Organization (ISRO). ISRO and Roscosmos for example have partnered for the mission Chandrayaan-2, which includes landing a spacecraft and deploying a rover by 2013 (ISRO, 2010). Two additional missions from China and Japan follow in 2013 and 2015 (Brown, 2010). NASA has also considered a GLXP-like mission with a robot that can be tele-operated from Earth and can transmit near-live video (NASA, 2010b). New space companies might also accelerate the development of space technologies. With support from NASA, companies such as SpaceX and Armadillo Aerospace for example are developing new launch vehicles and lander technologies.

Among the factors that are likely to affect the development of space technologies is the global economic slowdown which started in 2008. This unfavorable economic context has led to budget cuts, and space programs (both public and private) have been scaled down. NASA, the most important demand driver in this sector, has focused on supporting the development of technologies for commercial resupply linked to space stations and companies that can serve its needs. In the medium-term this and other space agencies will continue to be the driving force and induce a more stable demand for commercial services, but there is uncertainty about the programs that will be ultimately maintained (Futron Corporation, 2010a; ASAP, 2011). In the long-term NASA and other foreign agencies will continue to be the driving force, but private commercial demand will increase as well (Futron Corporation, 2010a). NASA's budget for lunar exploration is expected to grow in the long-term (CBO, 2004).

6.2.4   Technology Incentives

The proximity of the Moon has made it an obvious target for planetary exploration and a possible intermediate point to reach farther destinations. The natural resources of the Moon and their potential for the development of economic and scientific activities offer many possibilities (Schrunk et al., 2008). Robotic exploration will allow setting the initial stepping-stone for the exploitation of those resources and the development of human settlements.

A fair question is whether there actually is a significant market for the GLXP technologies and also to what extent it depends on other space programs and evolution of the space sector. When launching the competition the XPF expected NASA and other foreign space agencies to be the immediate near term market and GLXP teams to be able to provide data, heritage for new hardware, and at the broad extent risk management for government funded space missions. Teams were expected to eventually go into business 'flying lunar robotic missions for $50–60 million' (Hsu, 2010). Those predictions have not been realized yet, but there have been some market signals such as NASA's recent Innovative Lunar Demonstrations Data (ILDD) program with contracts awarded to six USA GLXP teams to purchase $30 million worth of data from commercial lunar missions.

Industry experts and some academic and private sources (and even GLXP team leaders, as described later) consider that there is a potential market for the GLXP technologies, but that their value and the time horizon for the realization of commercial opportunities are uncertain. This market would primarily comprise NASA and other space agencies, with private customers potentially increasing their share in the mid- or long-term. Experts consider that it is very difficult to know whether there is a market for commercial lunar surface activity of one kind or another and, therefore, the value of those potential market segments is uncertain. There is a potential government interest in planetary landers and Moon exploration technologies such as robotic remote controls for example (Hsu, 2010). But this market might depend greatly on the reactivation of more important NASA lunar exploration programs such as Constellation.

Futron Corporation, a technology management consulting firm, developed an estimate of the market value of lunar exploration technologies and presented it to the GLXP teams at the 4th annual GLXP Summit. This market may be worth between $1 billion and $1.6 billion for the next 10 years and includes market segments such as hardware sales for the government sector ($700 million), services for governments ($200–400 million), products for the commercial sector ($30–160 million), entertainment ($10–100 million), sponsorships ($50–100 million) and technology sales and licensing ($10–100 million) (Futron Corporation, 2010b). The study by Futron also notes the problem of assessing the size of individual markets due to the lack of a successful track record and the lack of comparisons with other markets. Other considerations include the government sector as the most important driver through the purchase of hardware and services. The private sector market for space hardware and other revenue streams may also be available to emerging companies in the sector. The GLXP teams may for example provide services of payload transportation, sell mission data or even provide some communication or entertainment services to private customers. NASA might pay between $4.5 million and $7 million per kilogram of lunar transportation (Futron Corporation, 2010a).

The market segments presented by that study represent different business and revenue models. Few data are available to assess opportunities in each case. For the case of payload services, another recent study by Futron Corporation about commercial lunar transportation suggests that '. . .the majority of investors view lunar transportation as a new, unproven industry without proven business models that provide multiple revenue streams' (Futron Corporation, 2010a). Therefore, the study continues, venture capital investors expect returns on investment of 40 to 50 per cent for space-related ventures. Private equity seeks returns of 30 per cent or more. To invest in space ventures, investors also require in the short-term strong commitments of NASA-funded programs, several 'beta-successful' companies and multiple revenue streams, among others (Futron Corporation, 2010a).

We shall consider that more generally the GLXP have been significantly affected by an adverse economic context. When the prize was launched in September 2007 the economic context was still favorable or neutral for the space industry. But about a year later the rumors of an important economic slowdown became more widespread and markets plummeted (the Dow Jones Industrial Average Index of the New York Stock Exchange went down about 25 per cent in the week after 22 September 2008). Since then the USA and the global economies have recovered fairly well but still with some uncertainty about the strength of this upswing. An adverse broader context for space activities may have affected other competitions as well. In the past decade the economic slowdown after the terrorist attack of

11 September 2001 and the increasing risk perceived in aerospace activities due to the loss of the space shuttle Columbia in 2003 affected perceptions and prospects of space activities related to the AXP (Maryniak, 2010).

It should be noted that commercial lunar exploration is not a new idea. This investigation was able to identify at least two private initiatives targeting such a market. The first is BlastOff!, a company founded in 1999 to develop entertainment space missions with a business plan based on sales of advertising, media content, merchandising and payload delivery.[21] The company planned $50 million missions featuring soft-landing, rovers for long distance travel (10–20 km) and HD video/image broadcasting (interestingly the company planned to use consumer-grade cameras for this) (Diamandis, 2008). The budgets for the first and second missions were $50 million and $20 million, respectively, and the expected revenues were at $250 million. Although the company was able to raise about $15 million in private funding it ceased operations after the dot-com stock market crisis of 2001 (Pomerantz, 2006). There was also LunaCorp, founded in 1989. This case is very similar to that of BlastOff! and is linked with the GLXP team Astrobotic.[22] LunaCorp planned to land a rover on the Moon and offered a number of related services such as tele-presence experiences and payload delivery. The parallel with team Astrobotic is very interesting if we consider that the projects were launched in very different industry and general contexts (Table 6.2). Both LunaCorp and Astrobotic have pursued similar market opportunities but with significantly different mission budgets ($250 million then, $100 million now). But while LunaCorp could not raise enough interest from sponsors and went out of business in 2003 (Reichhardt, 2008), Astrobotic has already received contracts from NASA and may have received funding from investors. This team was among the first to enter the GLXP and contributed significant technology outputs, as described later in this chapter. The team was also the first to publicly announce the signature of a launch contract for its mission with SpaceX.

6.3   THE PRIZE ENTRANTS

Thirty-five teams from 17 countries entered the GLXP before the closing of the registration period. Six teams had already withdrawn or merged and 29 remained in competition at the date of this analysis. Those 35 teams that entered officially include 17 USA teams and 18 other national or multi-national teams. The latter include teams such as Synergy Moon which reports members from at least 15 different countries (GLXP, 2010b). Several teams notably have members from countries that have not had any significant space program before, such as Bosnia, Serbia, Ireland

Table 6.2   *LunaCorp private initiative in the 1990s and GLXP team Astrobotic*

	LunaCorp (1989–2003)	Team Astrobotic (2007–present)
Approach	Two wheeled-rover missions	Wheeled-rover mission
Team	Business executives, scientists and former NASA officials; research from Carnegie Mellon University (CMU)	Spin-off of Carnegie Mellon University (CMU)
Budget	Varies, from $250 million (1996) to $80–200 million (1999)	GLXP mission will cost $90 million
Funding	NASA supports CMU's lunar robotics effort with $1.25 million a year	NASA's ILDD contract for $10 million (Nov 2010); NASA's two-year contract for $600 000 to develop lunar mining technology; NASA's purchase order for $500 000 (from a $10 million total) for hardware demonstration (Dec. 2010); at least one private investor.
Revenues	From payload and other services, expect up to $365 million	From payload and other services, GLXP mission would generate up to $24 million in prizes and $175 million in other revenues
Launch method	Reusable Roton vehicle by Rotary Rocket or Boeing Delta II	Signed contract with SpaceX to use Falcon 9 to launch spacecraft (Feb. 2011)

Note: This comparison should not be interpreted as an indication of the performance of any team; monetary figures in US dollars.

Sources: LunaCorp(1996), Cronin (2011), Bloomberg Business week(2011) and Astrobotic's website.

and Sri Lanka. This numerous and widespread participation exceeded the initial expectation of the XPF of about a dozen teams from a few countries (Pomerantz, 2010a). In fact many more potential entrants have demonstrated interest in this competition. In the first 18 months of competition the XPF received more than 2500 inquiries from individuals, companies and universities from 96 different countries (Pomerantz, 2011a).

To enter the competition the XPF required entrants to be some type of legal entity (e.g. company, foundation) and sign a Letter of Intent to Compete. Entrants also needed to submit an application package with diverse information about the team, its members, finances and mission plan, and had up to 90 days to formalize its participation (or lose the registration fee). Only those applications deemed serious were accepted. Non-serious potential entries comprised those applicants that were 'completely unaware of what they were getting into', only seeking access to the brand of the competition or its sponsors and partners, or critically reliant on demonstrably impossible methods (Pomerantz, 2011a). Although applications were not rejected in any case, between five and 10 applicants were invited to re-submit revised versions of their applications. Registration fees acted as an external validator at this stage. Fees were considerably lower compared with the expected cost of a GLXP mission. The XPF assumed that, if would-be entrants were not able to raise funding to cover the entry fees based on a credible project proposal, then either their intentions were not serious or their mission designs had significant flaws. The most significant difference between the GLXP teams and those that ultimately did not enter the competition might have been the level of seriousness of the applications. Inquiries that did not turn into formal entries seemed to be more 'spur of the moment' (Pomerantz, 2011a).

Interestingly most of the teams did not enter the competition until months or even years after the prize announcement (Figure 6.1). If we consider all the teams that have ever officially entered the GLXP, the breakdown by entry period indicates: 14 teams in the first year, eight teams in the second year, three teams in the third year and 10 teams after three years of competition just before the registration period closed (there were also a handful of teams that signed a letter of intent to compete but ultimately did not enter the prize). Unless otherwise indicated this analysis focuses on 17 teams that participated by responding to questionnaires, accepting interviews and/or allowing site visits to their workplaces. This investigation gathered team-level data until December 2010. To that moment only 26 teams had been officially announced as competitors and 23 of them were still active. The rest of this chapter further elaborates on this interesting phenomenon of late entries.

The GLXP entrants are generally newly created teams that adopt diverse legal forms. They include for-profit organizations (47 per cent of the teams), non-profits such as foundations (29 per cent) and independent, informally organized teams (18 per cent). Only one team is part of a larger organization, in this case a medium-size company. Some evidence indicates that the legal form a team adopts is likely to depend on the country the team is based in and the degree of progress of its project. Five out of eight for-profit teams (companies) reported to be based in the USA and four out of five non-profit

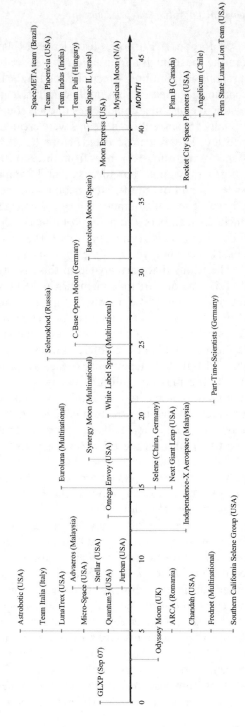

Note: The timeline indicates months since prize announcement; the deadline for registration to enter the competition was month 39.

Source: Based on official press releases by the XPF.

Figure 6.1 Timeline with official entry period of GLXP teams

teams reported to be based abroad. A European team leader explained in this regard that incorporating the team as a foundation was easier but also an impediment to raising significant funding, which later on led the team to explore the option of becoming a company. Some multi-national teams also comprise groups with distinctive organizational forms in each country for similar reasons. Operationally 65 per cent of the teams were created exclusively to enter this competition. The six teams created before the GLXP had previous working experience as a group that ranges from three to 20 years.

The GLXP teams are generally small but their size and composition vary significantly across teams and over time. New people join the teams and some people step down. The total number of people engaged in a typical GLXP project includes members and active volunteers. Members (the 'core team') are those individuals permanently with the team on a full-time or part-time basis according to the time they spend working on the GLXP project.[23] The teams that participated in this investigation had between one and 40 full-time or part-time members in 2010, with an average of four full-time and 11 part-time members. Smaller team sizes are generally explained by either a recent entry date or the early phase of project development. There are also collaborators and volunteers that work for the team sporadically, sometimes remotely, only when they are needed for specific tasks. All but three teams enrolled an average of 14 volunteers each by 2010, with a maximum of 80 volunteers for a team. Considering both members and volunteers, the author's estimate is that the 17 teams that participated in this investigation engaged at least 438 people in 2010 (this is about two years and a half after the prize announcement). The actual total participation in the GLXP is likely to be much higher. Teams such as FREDNET are evidence of that.[24] This team has pursued an open-source approach to the GLXP project and has enrolled more than 500 members and volunteers from about 40 different countries, and has likely been the biggest team (Evadot, 2009).

Membership estimates at any given time may be misleading in the analysis of prize participation because teams grow fast and significantly after they enter a competition. At the moment of their creation the size of the GLXP teams ranged from just a few volunteers working sporadically to 15 people working either full-time or part-time. Only two teams started with a significant number of volunteers (60 and 15, respectively) when created. Teams grew an average of 170 per cent between their entry date and December 2010. The most significant average growth was found in the number of full-time members (about 190 per cent), followed by part-time members (about 140 per cent) and volunteers (about 90 per cent).[25] There are a number of methods that teams use to attract and hire new members. Some teams successfully recruit new members at conferences and industry events. Team Part-Time Scientists for example grew from 40 members in April 2010 to about 70 in October 2010 by that method. Another method used by a few teams is giving presentations and talks at universities to engage students and sometimes meet new partners. Having an online presence also helps teams significantly in this recruiting process. This presence comprises not only a team website but also extensive use of social network platforms such as Facebook and Twitter.

Team membership is diverse and in a range of education levels. An average of 58 per cent of the members of each team has a background in engineering, 19 per cent in physics/chemistry/mathematics, 14 per cent in computer science/IT and 23 per cent in other backgrounds.[26] A significant proportion of members have achieved graduate education. Five teams have 90 per cent or more of their members at that level. An average of 15 per cent of the members of each team has reached the Ph.D. education level, 42 per cent the Master's level, 30 per cent the College/Bachelor level and 9 per cent reached only a High School level. Overall GLXP teams engaged about 100 people at the Master's level, 40 at the Ph.D. level and 80 at the College/Bachelor level during 2010. The prize also enabled students and women participation in a field traditionally occupied by a more experienced male workforce.[27] On average students represented 27 per cent of the members of all but two teams in 2010. Women represented 24 per cent. In total more than 66 students and 50 women were engaged as team members that year, and many more actively participated as volunteers. The leader of a 20-people, university-based GLXP team estimates that up to 200 students may have volunteered for his team during the first three years of competition. This student participation fluctuates considerably over time however, increasing more notably when students return to classrooms at the beginning of each semester. Their contribution to the GLXP projects varies as well. While some team leaders share their excitement in this regard, others consider that students require a significant supervision effort, which might offset the benefits of having them in the team. Students do not hide their enthusiasm. An engineering student that leads the development of the lunar lander of a GLXP team explains:

> . . .we're extremely passionate. We're right in the heart of our careers. We are learning a lot we know everything off of the top of our heads, and we have fresh thinking. We don't have any traditional designs that would influence the way we work. We can kind of come up with new approaches and new ideas and just run with it, and not have to worry too much about what we've done in the past because we haven't done much in the past. [. . .] We have all of our time to dedicate to these things, we don't have to focus on other projects that we are getting paid to do and put this aside. We can continue working on it. So, things are getting done pretty fast.[28]

Table 6.3   *Main characteristics of GLXP teams that participated in the investigation*

	Unconventional teams						
	T4	T6	T7	T11	T13	T16	T18
Created exclusively for GLXP?	Yes	Yes	Yes	Yes	No	No	Yes
Type of entity	Profit	Non-profit	Indep.	Profit	Non-profit	Non-profit	Part other[a]
Origin	USA	Foreign	Foreign	Foreign	Foreign	USA	Foreign
Members	20	10	20	40	38	11	15
Volunteers	8	5	50	10	80	n/a	0
Space agency -	5%	10%	0%	10%	0%	0%	13%
Industry	–	–	–	–	–	–	–
experience (% of members)	10%	0%	5%	20%	16%	0%	33%
Students (% of members)	71%	0%	0%	25%	18%	100%	27%
Predominant background (% members)	Engin. (45%)	Other (67%)	Engin. (80%)	Comp. Sci./IT (31%)	Engin. (53%)	Engin. (100%)	Engin. (67%)

Note:
Classification of teams into unconventional and conventional is based on significant proportion of members with experience in space agency/industry
a.   team is part of a larger organization.

Source: Questionnaire to GLXP teams.

Overall the work experience of the teams varies but is still somewhat linked to aerospace technology research and development. On average 34 per cent of the members of each team have aerospace industry experience and 15 per cent space agency experience. Based on this work experience this investigation classified teams into 'unconventional' and 'conventional' (Table 6.3). Eight teams with significant space agency/industry experience (i.e. 50 per cent or more of their members have that kind of experience) have been classified as conventional teams, or teams that are generally familiar with the development of space technologies. Nine other teams have been considered 'unconventional teams'.[29] There are other team members with related theoretical or practical knowledge as well. Some team members (about 30 per cent on average) have undertaken academic research in aerospace-related topics and others (about 20 per cent) have some type of rocketry experience as an independent professional.

Table 6.3 (continued)

Unconventional teams		Conventional teams							
T19	T23	T3	T14	T20	T21	T22	T24	T25	T26
Yes	Yes	Yes	Yes	Yes	No	Yes	No	No	No
Non-profit	Indep.	Profit	Indep.	Non-profit	Profit	Profit	Profit	Profit	Profit
Foreign	USA	USA	USA	Foreign	Foreign	Foreign	USA	Foreign	USA
6	1	21	12	1	12	7	11	4	2
25	3	0	2	15	2	6	0	n/a	1
0%	0%	24%	0%	100%	17%	57%	36%	0%	50%
–	–	–	–	–	–	–	–	–	–
0%	0%	71%	100%	100%	100%	86%	55%	100%	50%
67%	100%	24%	8%	0%	0%	29%	9%	n/a	0%
Engin. (67%)	Engin. (50%)	Engin. (71%)	Engin. (67%)	Engin. (100%)	Engin. (58%)	Engin. (57%)	Engin. (64%)	Engin. (50%)	Engin (100%)

Participation in other technology competitions is also a relevant experience that a few GLXP teams and about 7 per cent of all team members have. Team Micro-Space has participated in both the GLXP and the N-Prize. Team ARCA participated in the AXP. Team Rocket City Space Pioneers' leader was part of Scaled Composites, the team that won the AXP. Team Phoenicia also participated in the NGLLC. Some members of team Astrobotic participated in the DARPA Challenges and won the 2007 edition of the competition. A handful of team leaders also have some entrepreneurship experience prior to this competition.

The examination of the goals of the teams completes this overview of prize entrants. Sometimes team goals are reflected in their performance but in other cases team strategies and ultimate goals are less apparent, as further discussed in the next chapter. Teams may seek to demonstrate technological leadership for example and they take it very seriously ('When you lead a team, the world expects you to win'[30] explains a team leader). Other goals may be achieved with the mere participation in the competition. Some teams for example explain that they have entered the GLXP to learn and gain hands-on experience with space technologies, gain reputation and develop their professional networks or even inspire other people and countries to take an interest in space exploration. Others participate mainly for fun or for other things learned along the way. The XPF has also identified some GLXP teams that 'honestly know that they do not have high chances of winning the prize' but still participate for diverse reasons (Pomerantz, 2010a).

There are also a number of teams primarily focused on the creation of a commercial enterprise that do not necessarily contemplate Moon landing. The XPF's Director for Space Prizes explains:

> We have a number of teams that view their ultimate line of business primarily as a support business, so where they are going to develop the best mission control for a robotic lunar mission, and even if they don't get all the way to the moon, they're going to develop that step, and they're going to sell it to other teams, and they're going to sell it to Boeing, Lockheed Martin, and NASA, and aren't really anticipating getting to the further step unless something in the market changes or something in the company changes in a way that they don't foresee.[31]

Moreover, when asked specifically about their next step after the competition ends, all but two teams declared their interest in continuing their work on space development and commercializing the technologies they developed for the competition. By 2010 more than 60 per cent of the teams (including those with open-source approaches) had already appointed at least one person to focus exclusively on business development. Almost all of the teams will also seek to continue research in aerospace/aviation/satellite communications topics and only one team reported that its members will retire from this kind of career or project. One-third of the teams will also seek to enter other prizes and half of them have not yet considered the possibility.

6.4   MOTIVATIONS OF PRIZE ENTRANTS

The reasons to enter the GLXP are several and very diverse. There are four predominant motivations. First there is the opportunity to participate in a real technical and intellectual challenge. More than 80 per cent of the GLXP teams consider this an important motivation. The GLXP mission is a challenging project even for those teams with more resources and skills and is very stimulating for both more experienced professionals and students as well. Most importantly the GLXP represents an opportunity to learn and gain hands-on experience in aerospace technology development and project management. While team members get to work with cutting edge technologies and on diverse aspects of their projects from a system-level perspective, engineers in space agencies typically work on the development of specific components or subsystems for a larger project and do not deal with the entire spacecraft or project at the system-level (which sometimes becomes 'a bit frustrating', some GLXP engineers confess).[32] Team members also have the opportunity to work in an environment that is more competitive and exciting than traditional space agency or industry work settings. A few GLXP engineers also explain that the competition offers a concrete goal and a measure of personal achievement and technical ability that the space industry does not generally offer.

Second there is the potential commercial value of the technologies involved in the competition, which is an important motivation for at least 65 per cent of the GLXP teams. The potentially sizable market for lunar exploration technologies is not exclusively related with the GLXP and is available not only to prize entrants but also to other new and established industry players. We shall consider that the total $30 million in cash purse is a significant amount of money but does not compare with the sizable potential markets for the prize technologies. A GLXP team leader points out: '. . .the point is if one of the teams lands on the Moon, it will get much more than the $20 million. [. . .] If you have really a working technology, a reproducible working technology, those $20 million are not the point anymore.'[33] On the other hand while teams consider scenarios for lunar exploration technologies that do not differ significantly from those of industry experts and other private sources, in a handful of interviews they look much more confident when discussing commercial opportunities and the expected time of their realization.

Third, associated with the market value of the technologies, there is the opportunity the GLXP creates to gain recognition from NASA or other government agencies for potential future contracts. Almost half of the teams consider this a very important motivation. Although the competition is not officially sponsored by NASA or any other space agency, it generally exposes the teams to greater visibility and potentially captures the attention of program managers interested in similar technologies. In their market forecasts the teams confirm that they expect governments to be the main target in the short- and medium-term. The USA teams that seek to create a commercial enterprise expect to have NASA as the biggest single customer. A few teams also venture to anticipate growing private demand for services such as payload delivery and even Moon mining missions in the medium-term.

Finally there is the potential benefit that this kind of space exploration project can bring to society. At least 70 per cent of the GLXP entrants consider this an important motivation to enter the prize. This factor however is related with the very nature of this technological field and other idiosyncratic aspects at the individual-level and not with this competition in particular.

The GLXP has also created a number of other important incentives associated with opportunities generally available to all entrants but their actual perception depends on the team and its goals. About one-fourth of the GLXP teams for example are particularly attracted by diverse resources they get access to when they enter the prize and the competition develops. Investors, sponsors and other individuals and organizations interested in the GLXP projects approach the teams and facilitate access to funding, in-kind resources and volunteer effort for the GLXP projects and also other projects the teams have. The funding requirements of this kind of mission make this opportunity to gather resources particularly relevant in this competition. A few team leaders also remark on the prize sponsorship value created by the Google Inc. brand, which increases the credibility of their projects when they seek funding or team sponsorships. That relates with another valued opportunity offered by the prize to expand professional networks to access knowledge and other resources. Partnerships, fundraising efforts and events such as conferences, GLXP annual meetings and industry fairs help teams to develop valuable networks that can transcend this competition.

Only one-fourth of the GLXP teams consider the cash purse a very important motivation. Another 40 per cent of the teams consider it somewhat important and a handful of them indicate the cash purse is not important at all. The evidence suggests that the interest in the cash purse might be related with the strategies and ultimate goals of the teams. The prize money might eventually become part of the payoff promised to investors, another source of income in a longer-term business plan or simply a reward for the team members. It should be noted that the GLXP also offers monetary rewards that are not linked to the performance of the teams. The $1 million Diversity Award is for the team that, in the opinion of a panel of experts appointed by the XPF, makes the greatest attempts to promote diversity in the fields of science, technology, engineering and mathematics (STEM). Although this amount of money is considerably lower than the grand prize, it may be still a significant source of income or reward for teams interested in that kind of goals such as the USA team Jurban, a non-profit team focused on motivating underrepresented students to enter STEM fields as they apply to space entrepreneurship.

There are other diverse reasons to enter the prize that only a few teams consider important. A few teams are particularly motivated by the opportunity to build reputation and demonstrate their leadership in the field. Prize teams can get a lot more publicity than comparable non-prize projects because prize sponsors also publicize the efforts of participants and a word of mouth effect further spreads information about the competition and its competitors through the traditional media and online social network platforms. Teams can astutely use this effect in their favor, a team leader suggests: '. . .teams that are slow respond later. So when Google announces the prize, or the foundation announces a prize, they think it is a story about themselves and the prize. But if you are quick, it is a world story about your team.'[34] The teams and members with good performance in the competition can not only build business, professional and personal reputation but also demonstrate technological leadership. The prestige and publicity associated with accomplishments in the competition also have a strategic value for those teams that seek to enter markets or position themselves in the academic or professional arena. Two other teams indicate that the GLXP provides additional incentives to reach other goals beyond the prize. The prize, team engineers explain in interviews, focuses their efforts, gives them something to compete against, makes them push their limits and work faster, and sets a time line to organize their work. Another team has been very motivated by the opportunity to demonstrate the viability of a technological concept using the GLXP mission as a test case. Most of the members of this self-funded team worked together for many years in the space industry and considered that a concept developed by them years ago was technically appropriate to accomplish this mission. Finally a few other teams have been also motivated by the recognition they might gain from family, friends and colleagues, the entertainment value of the competition and even religious reasons to participate in technology development.

Late entries suggest that motivations to enter the GLXP may have changed over time. The analysis of a set of motivations probed in questionnaires shows notable differences in motivations between types of teams (i.e. unconventional and conventional) and along different entry points (Figure 6.2). According to their official entry date, only 40 per cent of the teams that participated in this investigation (three unconventional, four conventional) entered the GLXP officially in the first year of competition. The rest of the teams (six unconventional, four conventional) entered in the second and third years of competition. Unconventional teams are generally driven by prize incentives and only to some extent by technology incentives, but this pattern changes over time (Figure 6.2a). The unconventional teams that entered the competition in the first year primarily emphasize the benefits the project may bring to society, and also consider learning and commercialization of prize technologies as important motivations (Figure 6.2a). The teams that entered during the second and third years of competition, however, primarily emphasize the importance of participation in a real technical challenge. Learning and

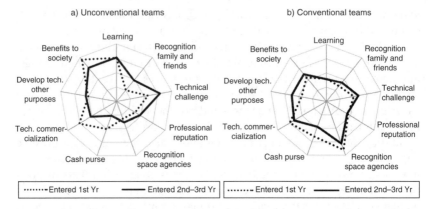

Note: N=17 cases; radar charts show importance given to each type of incentive, by type of team and prize period; number of teams first year=7 (3 unconventional, 4 conventional); number of teams 2nd and 3rd years=10 (6 unconventional, 4 conventional).

Source: Questionnaire to GLXP teams.

Figure 6.2   Motivations of GLXP teams, by type of entrant and entry period

potential benefits to society are also important for this late group but to a lesser extent. Developing technologies with a different, non-prize purpose is the next most important motivation for this second cohort (more important than commercialization of technologies). The importance of the cash purse is notably lower for unconventional teams that entered the competition after the first year. On the other hand conventional teams are remarkably driven by the opportunity to get recognition from NASA and other space agencies (Figure 6.2b). The pattern of motivations does not change very much over time and the potential value of the prize technologies remains as a strong incentive in both time periods as well. Conventional teams that entered the GLXP after the first year of competition, however, consider the development of technologies for other purposes and the participation in a real technical challenge more important. Benefits that technology development brings to society are also relatively important for conventional entrants. Not surprisingly, learning is not a motivation for them (by definition this group has space agency/industry experience).

The interviews with some GLXP teams confirmed what preliminary analyses of documentary data anticipated. The teams generally involve very proactive, intrinsically motivated individuals that are driven by diverse interests. This is possibly best reflected in expressions of the kind 'this [prize] has my name on it' or 'we can do it, let's do it' that team leaders use when they describe their reasons to participate.[35] These interests may be even conflicting from the point of view of the competition, which can be clearly perceived for example in relation with the importance given to the cash purse and potential commercial opportunities available to the team (team members not interested in monetary rewards might prefer designs that are not commercially viable). Moreover, while some team members are fully committed to the GLXP project and strongly aspire to claim victory, other members find satisfaction in the accomplishment of more focused tasks or in taking part in the development of specific mission components. This is also related with broader perceptions and ultimate goals of each individual. A team leader notes in this regard: '. . .we have people who believe in the entire mission and we have people who believe in their single part of the mission.'[36] Intrinsic motivation is a pervasive underlying factor that explains prize participation. The author was able to interview very proactive and persevering people that enjoy their participation in the prize very much, although sometimes this participation requires dealing with obstacles or lack of support at the personal level as well.[37]

Whether the prize teams are less risk averse than traditional industry players (as some literature suggests) is another facet of their motivations and reasons to enter the competition. In the pursuit of a GLXP-like project there are not only the technical risks associated with the development of space technologies but also other risks emerging from prize participation such as those related with excessive financial exposure, cash purses not paid, or wasting time and effort. The perception of these potential risks is certainly influenced by a number of factors including the form of organization of the team and the knowledge and skills of its members. This perception may change over time as well, as teams make progress in their projects. While teams that are in early stages of development may consider that '. . .there is not much you can lose, except your time',[38] front-runners are likely to have much more at stake.

When asked in questionnaires and interviews about this, however, the teams respond that there are no significant risks associated with their participation in the GLXP. They consider that the two most important risks they might face are taking on excessive financial commitments and compromising other activities or projects of the team members. Only five teams indicate those as major concerns, and the rest of the teams worry a little or nothing about those potential risks. All but a handful of the teams consider that losing invested time and resources if they ultimately fail to win and embarking on ineffective technological approaches are not risks at all. And whether the prize sponsor ultimately pays the cash purse if they complete the mission first is not a major concern either. Unconventional teams in particular are even less risk averse as about 70 per cent of them consider that those factors mentioned above are no risk at all. Only about one-fourth of the unconventional teams are concerned a little with those risk sources. Financial risks are still their most important concern but to a lesser degree compared to the rest of the teams. On the other hand about 40 per cent of the conventional teams are particularly concerned with compromising other activities of the team members or taking on excessive financial commitments. In the end, due to the particular characteristics of the GLXP, there are always some financial, business and legal risks that result from, for example, sponsorships or partnerships, and those risks affect teams to different extents depending on their form of organization. As described in the next sections, teams seek partnerships to commercialize their technologies or to source technologies in exchange for IP rights, percentage of revenues or other benefits. These partnerships imply either a risk that has to be mitigated by reducing the team's commitments or an opportunity to actually decrease the risks by pursuing a much more collaborative effort.

On the other hand the team leaders are aware of the technical risks involved in the GLXP mission and emphasize those in the testing and launching phases, but promptly clarify that those are not prize-specific risks. A few leaders also indicate that their teams apply risk management procedures, which one of them described as 'pretty similar to the one that NASA has'.[39] Risk considerations in the GLXP involve both R&D risks and the so called programmatic risks. The former are inherent to any R&D process and are related with the cost, schedule and technical performance of the technology. The advancement of technologies with lower maturity levels is particularly associated with higher R&D risks. The programmatic or mission risks are related with the uncertainty about whether the mission will actually fly. It is important to consider that a mission with a catastrophic loss has no revenue and therefore it would imply a very significant economic cost for a team that has reached the point of a launch.

6.5   PRIZE R&D ACTIVITIES

6.5.1   Design Criteria and Idea Sourcing

The GLXP teams use very diverse approaches in their projects and there is no design criterion that predominates in all the projects. Considering the seven criteria probed by this investigation, technical simplicity and project cost are the most prioritized, with about 40 per cent of the teams assigning them the maximum priority and 75 per cent or more of the teams considering them among the top three criteria in project designs. The market value of the technologies, the next in importance, is the top criterion for 20 per cent of the teams and in the top three for about 45 per cent of the teams. Novelty is low-ranked among these seven design criteria. It is the top criterion for only two teams. Environmental impact and standardization are the least considered design criteria. There are no significant differences between the design criteria used by different types of teams, except for the marked use of technical simplicity as the top criterion by unconventional teams. While 67 per cent of these teams prioritize simplicity, only 14 per cent of conventional teams do so. Conventional teams are more likely to consider project cost as the main design criterion, and market value, novelty and technical simplicity among the second or third most important design criteria.

Some team leaders suggested additional design criteria and insights on criteria that are specific to the prize context and/or represent innovative approaches to aerospace design. The three most frequently cited criteria are reusability (i.e. systems that can be used for multiple missions), optimization (i.e. systems that balance operational efficiency and performance with prize challenge achievement) and performance (i.e. systems that meet the mission's requirements and minimize failure and maintenance). Other less common design criteria include 'simple and smart' designs (i.e. creative, simple solutions that work effectively) and 'minimalism' (i.e. capabilities that stick to the minimum prize requirements). Interestingly a few teams also mentioned 'scalability', a design criterion that is commonly used in telecommunications and software design, which refers to systems that can be scaled up to meet requirements of larger missions. Some of these design criteria are exclusive to unconventional teams. These include performance, minimalism, robustness (i.e. capacity to resist many mission days) and scalability. Finally, there is the minimum technology development effort, a design criterion exclusive to the prize competitive environment that teams refer to as the design that 'gets you from A to B as fast as possible'. The words of a GLXP team leader are illustrative about the differences between this approach and practices in traditional government-led missions:

> Government missions can be a bit more expensive because you've got other reasons for spending that money. [. . .] We can't play that game for minimum cost solutions, and that impacts all aspects of the mission architecture. Also, we have a race so we have to have minimum development time. So these two points together mean that we want to have absolute minimum technology development effort. We don't want to do fancy things; we want to do simple things in a smart way.[40]

Further analysis of the designs of GLXP teams shows that they are mainly based off the team members' knowledge, available commercial products and past projects of the team members, with the major part of the teams considering them important or very important sources. One-fourth of the teams also draw significantly upon ideas found in non-aerospace projects and designs of teams participating in other prizes (in particular some technologies and ideas developed for the NGLLC might be useful for the GLXP; at least one USA team mentioned contacts with Armadillo Aerospace and Masten Space Systems, the winners of the NGLLC, and another foreign team mentioned unsuccessful intents to establish those contacts). The designs of other GLXP teams are considered the least important sources of ideas. Seventy per cent of the teams consider that designs of other teams in this or other prizes are not important at all. Unconventional teams have slightly different sources of inspiration compared to other teams. They find more inspiration in projects that space agencies have developed and designs found in non-aerospace projects. Their designs are also less based off previous projects of the team, which is related with the fact that most of these teams are newly created to enter the competition and, by definition, do not have industry experience. These teams may be also learning from other GLXP teams and teams that participated in other projects. Nonetheless unconventional teams also bring new approaches to space development. A GLXP team leader refers to this explicitly: '. . .the fact that our team isn't normally working on the subject and maybe we have somewhat like an outsider's perspective to this. So, we are looking at things differently like people who are doing this in an all day job.'[41] On the other hand conventional teams draw slightly more on ideas that come up from their theoretical knowledge and existing commercial products.

Teams also mention alternative design sources such as external expert advice, partners and online documentation. The use of external experts was mentioned by four out of seven interviewed teams. Those external experts are either part of professional networks the team members had from past projects or new contacts the team has developed to work on this specific project. Access to multi-disciplinary advice is also available to teams that partner with universities. The use of partners as a knowledge source for design was mentioned by only one team that strategically uses its network of partners to source technology and components for its project. Another team mentioned documentation published online, which comprises the work produced by other organizations that is freely available on the Internet, including work by rocketry clubs, declassified aerospace agency documentation and software tools provided by manufacturers, among others.

6.5.2   Own Development v. Use of Existing Technologies

Intuition suggests that there are a number of reasons for prize entrants to consider using as much existing technology as possible in their projects, which would help entrants to accomplish the prize challenge faster than their competitors. This investigation anticipated that teams might also use existing technologies to reduce project costs, facilitate commercialization of technologies and increase technology reliability in their projects. Technologies maybe readily available COTS (e.g. parts or components) or built after teams order them from a catalogue (e.g. solid rocket motors), but some testing and adaptation efforts are generally required for them to be used in GLXP projects.[42] This also applies to potentially useful non-space components and surplus parts that teams may adapt from government space programs. Effectively nearly 56 per cent of the teams indicated that the most important reason to use existing technology is reducing project costs. Commercially available parts and components can be very expensive but still cheaper than those developed from scratch. In other cases 'friend companies' or partners contribute parts or components to the projects in exchange for some value, or simply to have the opportunity to test their technologies in a space project. Thirty-eight per cent of the teams indicated they use existing technologies to increase the reliability of the projects, as the use of proven commercial solutions can reduce risks. Only two teams indicated that speeding up technology development is the top reason for sourcing existing technologies, but eight teams (50 per cent) ranked this reason as the second most important.

There may be other diverse reasons to use existing technologies. Most importantly, the lack of knowledge or expertise forces teams to rely upon technologies developed by specialized machine shops or experienced industry players. This leads some teams to partner with specialists in key subject areas to delegate the development of certain subsystems or parts. There are a few cases of teams that adopt a collaborative organizational approach in which a network of partners provide technologies and components that are integrated into GLXP projects. In some cases teams are bound by agreements to use certain technologies provided by partners even when that implies additional development or adaptation efforts. In other instances partners provide cutting edge technologies that teams would not have access to if they were not engaged in this project. For example, the German carbon-fiber manufacturer Crosslink-Fibertech provides Formula One-grade technologies to team Part-Time Scientists (PTS) in exchange for cooperation to open new markets (PTS, 2011a).

On the other hand there are reasons to develop own technologies

Table 6.4   Number of GLXP teams that subcontract and use COTS technologies, by type of entrant

Type of team	Subcontract these proportions of their project				Use these proportions of COTS technologies in their project			
	More than 50%	20% to 50%	Less than 20%	0%	More than 50%	20% to 50%	Less than 20%	0%
Unconventional	–	4	3	1	2	4	2	–
Conventional	2	3	2	–	1	5	1	–
Total	2	7	5	1	3	9	3	–

Note: N=15 cases (8 unconventional teams, 7 conventional teams); cells show number of teams for each range of subcontracting.

Source: Questionnaire to GLXP teams.

instead of acquiring existing ones. A handful of teams seek to develop their own technologies when technology commercialization or other projects are the focus of their activities, or when they have other organizational goals. A team that is very interested in gaining hands-on experience for example has sought to develop own technologies whenever possible. An engineer explains: 'We're trying to find a happy medium where we're developing as much technology as we can to get to the moon, but at the same time, we want to try to be able to compete in the timeframe that we are given.'[43] Finally ITAR regulations block foreign teams from obtaining some key technologies for their projects from USA companies, including propulsion, communications and navigation and control subsystems.[44] Teams without access to those technologies depend upon in-house capabilities or alternative non-USA sources (when available).

The extent to which teams use existing technologies or delegate development efforts to contractors rather than developing in-house is significant. All teams plan to use COTS components and parts and all but one team plan to subcontract part of their projects. More than half of the teams estimate that their systems will be somewhere between 20 and 50 per cent subcontracted to third-party developers and/or COTS technologies (Table 6.4). Two teams even indicate that more than 50 per cent of their systems will be subcontracted and three teams that more than 50 per cent of their systems will be COTS. But there are a number of teams making a more significant development effort. Five teams indicated that less than 20 per cent of their systems will be subcontracted and three teams that less than 20 per cent of their systems will be COTS. Unconventional teams might be less likely to subcontract their development efforts to third parties, but they still seek to reduce their engineering effort when parts/components are available COTS, from partners or friend companies. The data show that at least two teams with vast industry/agency experience entered the GLXP with intentions to subcontract most of their projects to either a set of partner companies or a single company.

6.5.3   Organization of R&D Activities

The forms of R&D organization that the GLXP teams adopt for their projects are very diverse but share some common features. The teams are generally flat and flexible organizations led by a small number of core members. Their R&D activities are geographically spread out and distributed across different locations and sometimes different countries. Within the same team there are workgroups that for example focus on the development of the lunar lander and groups focused on the development of the rover or camera subsystems. Four configurations of R&D activities were probed by this investigation to better understand the extent of this widespread activity. Seven teams (44 per cent) indicate that they organize their activities as different workgroups that work on the project from different locations; five teams (31 per cent) indicate that their members work remotely and only meet for some tasks; two teams (19 per cent) have different workgroups and regularly meet in the same location to work on the project; and only one team is configured as a single workgroup that regularly meets in same location.

The processes of technology development of these organizations are characterized by trial and error iterations, rapid prototyping and intense information exchanges. Unconventional teams seem more likely to show these features. A team leader provides a very illustrative description of what he calls the 'craft culture' of his team and how that relates with the cost and degree of achievement of project milestones:

> So, it's a craft culture. To get something like that done [he points toward a piece of equipment] might be $100 000. A group that has these values and experiences, and resources and facility [he refers to his team] might get the same thing done for $4000. But the money isn't the key thing. My point is there is also a tremendous inefficiency in getting it done in that traditional way, where there is the idea of what is needed which is transferred to a designer that puts it into some tangible form, which moves it to analysis, which determines if it is going to be this or that, that it should be changed in this way, which then sends it to a productions shop, which then orders the materials that then gets the things

done, which then goes to the assembly, which then goes to an inspection, which sends it back with communication and bills and all that kind of thing. It's very common around here to conceive something that is needed at this time, at lunchtime one day and have that thing, just like that, the next day. And, people die for it. They are all nighters. Three o'clock in the morning. People that could not or would not ordinarily be an analyst that have tremendous craft skills. They're what matter.[45]

There are also significant knowledge flows between the GLXP teams and other individuals and organizations external to the prize such as academic researchers, family and friends, consultants, contractors, colleagues with prize experience and even other GLXP teams. All but one conventional team with vast aerospace industry experience exchange information regularly with at least three of those types of individuals/organizations and three teams exchange information with all of them. More specifically, about 90 per cent of the teams exchange information with academic researchers, about 80 per cent of the teams do so with providers or contractors, 75 per cent with family and friends, about 70 per cent with consultants, about 55 per cent with colleagues with prize experience (e.g. former NGLLC competitors) and about 45 per cent with other GLXP teams. This investigation probed four main topics of information exchange. The most important topic of information exchange is solutions to technical problems for 75 per cent of the teams. About 30 per cent of the teams consider commercial opportunities the most important topic. Overall strategies to win the prize and team contribution to industry or society are the least important topics, with about 10 per cent of the teams considering them the most important.

Although some GLXP teams do exchange information with their competitors, insights from the 4th GLXP Annual Summit and site visits suggest sporadic interactions rather than stable or formal collaborations between teams.[46] Team leaders and representatives of only 11 teams (including some of the most active teams in terms of technology outputs) attended the Summit to present and openly discuss technical aspects and progress of their projects. Interestingly these presentations were not a requirement of the XPF but a suggestion made by a few teams after the first GLXP Summit. The author also had the opportunity to assist at a GLXP team's regular meeting, in which one of the members specialized in computer visualizations commented on his interactions with another team's member specialized in the same field and concluded that those interactions 'are making each other better'. Teams are generally open to discuss and share information about project developments on their websites and other online platforms, which are sometimes important sources of feedback from the public. Teams such as White Label Space and Team

Italia have even presented conference papers describing their GLXP missions.

To illustrate the differences across teams and their diverse forms of R&D organization, this investigation identified four types of exemplar organizations (Table 6.5). The Space Agency Legacy team is a conventional team with strong space agency/industry experience. It has a main corporate partner and draws upon an extended network of industry contacts. This team is organized as a handful of subgroups that work on the project from different locations including a few international groups. The Non-profit Partnership team is an unconventional team formed as a partnership between two universities and a newly formed foundation. It is led by aerospace engineering students. The Partnerships Network team is an unconventional team created as an independent team but has built significant university and corporate partnership networks. It has multidisciplinary background but predominance of computer science/ IT experience. Finally there is the University Spin-off team, an unconventional, university-based team with strong academic multi-disciplinary background, entrepreneurial experience and corporate support.

These teams approach the GLXP challenge in very different ways. For example, while two of these teams have core members that oversee all parts of the project and 'know it all', the other two teams have a more decentralized leadership because the project has become more complex (Space Agency Legacy) or because in that way 'you waste the cycles of all of your people, all of the time' (University Spin-off). The Space Legacy team has sought to implement more formal procedures and communications, documenting project tasks and using standards. This enables international collaborations and the organization of an enterprise with goals beyond the GLXP mission. Although team members explain that the team is much less bureaucratic than aerospace companies, project management has become more complex and looks more dependent on external funding to make progress. The other three teams (Partnerships Network, University Spin-off and Non-profit Partnership) quickly moved to prototyping and testing of solutions or early versions of their subsystems. These approaches are still systematic yet not bureaucratic. Although these teams do not document all their procedures, their tests inform further steps in technology development. The agility gained with informality has been paid with coordination issues and the need for internal re-organization in some instances. The Partnerships Network team in particular has engaged a few key members with aerospace experience and its internal organization evolves over time as the team learns how to better approach problem solving after iterations or cycles of development. The University Spin-off team also draws upon very experienced, multidisciplinary members and

Table 6.5 Organization of R&D activities of GLXP teams (exemplar forms)

Characteristic	Type of team organization			
	Space Agency Legacy	Non-profit Partnership	Partnerships Network	University Spin-off
Legal form/ type of team	Non-profit / conventional	Non-profit / unconventional	For-profit / unconventional	For-profit / unconventional
Priority goals	Professional reputation, publicity	Learning, other organizational goals	Pursue a challenging project	Demonstrate leadership, commercialization
Members/ volunteers	2 / 15	5 / 25 (est.)	40 / 10	20 / 8
Background/ experience	Agency and corporate aerospace experience	Aerospace students leadership	Multidisciplinary, computer science/IT predominance	Academic multi-disciplinary, entrepreneurial experience
Linkages/ partnerships	Corporate main partner; network of space agency/industry contacts	Foundation and university partners; university collaborations	University collaborations and network of corporate partnerships	Corporate and leading university support
Internal organization	Workgroups in different locations (incl. international members)	Workgroups in different locations	Mostly remote work (incl. international members), meet only for specific tasks	Co-location, 'everything under the same roof'

Key R&D features	• Industry standard procedures • Formal communications and face-to-face interactions • Entire project is not known to all team members • R&D dependent on external funding	• Rapid prototyping and testing • 'Cost effective organization' • Small core team that 'knows it all' • Access to specialized facilities • NASA-like risk management procedures	• Rapid prototyping and testing • Optimized workflow for idea sourcing • Low cost development structure • Agile organization • Open knowledge sharing • Small core team, key 'know it all' members at the center of the network	• Face-to-face communications • Multi-disciplinary inter-departmental collaborations • 'Craft culture' • Trial and error, simulation supports process • Iterative prototyping and testing cycles • Creative problem-solving when need to adapt technologies
Team example	T20	T16	T11	T4

Note: Characterization of **R&D** activities based on assessment of the author and descriptions of interviewees; not all teams define themselves in terms of each dimension—only features described by the interviewees are mentioned in the table.

Source: Questionnaires, interviews and site visits to GLXP teams.

Table 6.6   Number and type of partners of GLXP teams, by type of entrant

Type of team	Type and number of partners				
	Large company	SME	University	NGO	Total
Unconventional	5	22	5	3	35
Conventional	5	21	13	10	49
All teams	10	43	18	13	84

Note: Data as of January 2011, including 26 GLXP teams; based on teams that report at least one partner (i.e. seven unconventional teams; eight conventional teams.)

Sources: GLXP team websites and official press releases.

external advisors and, most importantly, past successful prize experience of key team members.

The GLXP activities span across the boundaries of the competition and involve multiple external actors (i.e. beyond the official teams). This occurs through different kinds of relationships between teams and other entities disseminated all over the world. Formal relationships generally include longer-term agreements with partners and sponsors to access knowledge and other diverse resources. In the first three years of competition, for example, at least 15 out of 26 GLXP teams have partnered with multiple other organizations to source technologies, seek support and advice and gather other types of resources needed to pursue their projects. Relationships with subcontractors, key technology sources, may also be included in this group. Informal linkages include indirect access to resources through volunteers and their organizations and other sporadic collaborations. These linkages comprise the volunteer effort described in prior sections and also what teams refer to as friend companies that contribute parts and components to their projects.

Partnerships are among the most distinctive features of prize R&D activities and the evidence of the widespread effect of the GLXP. In total the GLXP teams have engaged more than 80 partners in three years of competition, including 10 large companies, 43 SMEs, 18 universities and 12 NGOs (Table 6.6).[47] Companies become partners of teams (e.g. Sierra Nevada Corporation, partner of the USA team Next Giant Leap) or, in a few cases, team up with other organizations to form new teams (e.g. Dynetics, a privately held USA company with defense technologies expertise that is new to the space business, teamed up with Teledyne and Andrews Space to form the team Rocket City Space Pioneers). At least 12 teams also partner with universities and NGOs.[48] There is also at least one instance of collaboration with government agencies.[49] The number and types of partnerships vary across the GLXP teams. Conventional teams have partnered with more organizations than unconventional teams in the first three years of competition. They have also been more likely to partner with universities and NGOs than other teams. Unconventional and conventional teams are similarly linked to both large companies and SMEs, yet the median number of corporate partners by team is larger for conventional (median=4) than for unconventional teams (median=2).

6.5.4 R&D Effort

Equating the GLXP teams' R&D effort only with the total mission cost would be misleading, but estimates on the latter help to better understand the nature of that effort. Before the announcement of the prize, Marshall et al. (2007) suggested that a small unmanned lunar lander mission could be accomplished in under 36 months and with a total mission cost of about $88 million. The XPF estimated that each GLXP team would invest between $15 million and $100 million, with an average cost of $60 million, to accomplish this challenge in no more than 60 months (XPF, 2008b; Pomerantz, 2011a). Interview data and team websites show estimates that range from $4 million to $90 million, with only four teams that expect to accomplish the mission with $30 million or less.[50] A simple comparison of these estimates with the historical data on robotic planetary missions shows that the GLXP projects tend to be less expensive but require much longer development times than past missions, resembling the typical trade-off between cost and time in space projects (Figure 6.3).

To illustrate the R&D effort involved in a GLXP mission we can consider a simplified scheme of a $30 million mission based on discussions with team leaders (Figure 6.4).[51] There are six phases that represent increasing degrees of progress and final achievement of the mission. Phase 0 or initial phase represents the team's entry and development of initial ideas, and its cost relates with the competition entry fee and initial team formation. Phase A involves the development of initial models and mockups of spacecraft that teams use not only for mission accomplishment purposes but also to raise funding and seek sponsorship opportunities. Teams grow significantly in this stage. Phase B comprises further development of prototypes, testing and initial production of subsystems. Teams also use their prototypes for public demonstrations that give them access to new members and increasing visibility. Phase C comprises the production of subsystems and further testing. Phase D continues with production at the system-level and system-level tests. Finally Phase E comprises final preparations, launch and mission accomplishment.

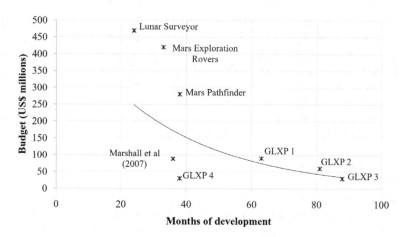

Note: GLXP 1–4 represent lead time/budget estimates of four different teams for GLXP mission achievement; exponential trend line added for interpretation.

Source: Author's analysis based on data from sources cited in the text and team interviews.

Figure 6.3   Lead time/budget comparison for selected past robotic missions, literature and GLXP projects

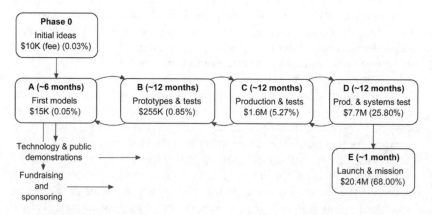

Note: The description of phases is only illustrative and does not seek to represent the budgets and processes of all teams; duration of iterations may vary and/or overlap; monetary figures in US dollars.

Source: Interviews with GLXP teams and other sources cited in text.

Figure 6.4   Phases and cost estimates of an illustrative $30 million GLXP mission

Although the costs shown in this scheme do not necessarily represent cash flows and the duration of phases may vary significantly across teams, the illustration is insightful about the financial gaps that any team might face. There is almost no correlation between how much gets spent and what really gets accomplished in this type of project. The most significant engineering effort is made during initial phases (phases A through C in this illustration) to develop systems such as rovers and landers, which account for about 6 per cent (or about $1.8 million) of the total mission costs in the first 30 months of the project. The production of the systems that the team will ultimately use in its mission and further testing stages (phase D) also require engineering effort and can represent about one-fourth (or almost $8 million) of the total mission cost. Concrete examples are helpful. Initial prototypes can be produced quickly and at relatively low cost. Team FREDNET developed for example a handful of small rover prototypes for concept testing purposes within its first year of competition. One of them, a small ball rover, was built for only $1000 (XPF, 2009). Team Part-Time Scientists, within its first 16 months of competition, built a more complex prototype that demanded about six months of development and costs between $41 000 and $55 000 including all the mechanical parts, the electronics and most of the other required subsystems but excluding labor and an 'unbelievably expensive' solar panel antenna introduced by the team (PTS, 2011b). Production and further testing stages are however more expensive. Team Odyssey Moon for example estimates the cost of its lander (developed under a Space Act Agreement with NASA) at around $3.5 million including $500 000 in parts and $3 million in labor (Kay, 2010).

Notably about 70 per cent of the total cost of the GLXP mission is still with the final phase of the mission, which includes the launching vehicle as the single most expensive item of the entire project. Commercially available launching rockets may cost between $10 million and $50 million depending on payload capabilities.[52] Rockets with smaller payload capabilities are less expensive but in turn require the team to exert increasingly significant engineering efforts to develop smaller and lighter spacecraft. On the other hand rockets with larger payload capability allow teams to reduce their engineering effort and use cheaper components. Teams do not have to afford this cost until about 12 months before launch, a period that in fact can be negotiated with the launch provider who ultimately has to build and integrate the payload after the contract is signed. Launch costs might be significantly lower if a team develops its own launcher, but cash flows would be much more significant earlier in the mission timeline. On the other hand the GLXP engineers consulted about this suggest that it would be impossible to accomplish such a thing within the development time frame given by the prize.

Interestingly while there are some teams that generally follow this illustrative path and seek to optimize their efforts to accomplish the prize challenge faster than others, there are other teams that design, test and develop technologies that may be useful for the GLXP mission but also involve significantly larger efforts and increasing development lead times. A GLXP engineer explains the optimization approach in these terms: '...making something last five hours is significantly easier than making something that can last years, and even with a small budget and a very hard time constraint, if you really want to focus on this one goal, then you can ignore lots of stuff'.[53] But there are on the other hand many examples of efforts that aim at much more than the GLXP minimum requirements. There is for instance the IAR-111 carrier airplane developed by team ARCA which will be able to take off from sea and to transport another vehicle (a launcher also developed by the team) to a certain altitude for ignition and further travel to reach the Earth orbit (XPF, 2010). In other words this team, rather than using commercial options, has chosen a path that others consider 'impossible' in this prize.

The lack of up front funding for their projects induces the teams to engage in fundraising activities. In the GLXP, teams have generally raised funding from team sponsors or generated income from technology commercialization. Only a few teams have been able to successfully attract investors. Sponsors contribute both monetary and in-kind resources such as components, parts or at least a discount in the team's purchases. In exchange the teams offer publicity and promotional actions that include for example placement of sponsor logos on prototypes or actual spacecraft, acknowledgements and credits in conference presentations or websites, or even the opportunity to name the team.[54] Some teams have also been able to market the technologies they develop for the prize. NASA's ILDD contracts to six teams are an example of this. Another common source of income is the commercialization of payload services, i.e. teams offer customers the delivery of cargo to the Moon. This is an attractive opportunity for universities and companies to send small instruments to the lunar surface at a generally low price tag per kilogram. For some GLXP teams this is a key component of their business model because it allows them to generate income before mission launch (agreements typically require customers to pay in advance). Other teams find that some of their technologies, existing or newly developed but not aimed at commercialization, have some market demand and use the opportunity to cash in and fund their GLXP missions. Team Phoenicia for example expects to cover between 15 and 20 per cent of its total mission cost with that income (Team Phoenicia, 2011). Finally only a handful of GLXP teams have received private investments according to publicly available data. For instance Team Next Giant

Leap raised funding from two private companies ($225 000); eSpace: The Center for Space Entrepreneurship (more than $30 000) as well as its own founder ($200 000). Draper Laboratory, one of the team's partners, also committed over $1 million from its internal R&D program to fund the design and development of a guidance, navigation and control system testbed for use in the team's mission (Kolodny, 2011). Another example is Team Astrobotic, which seeks to raise at least $25 million from private investors and cover the rest of the mission cost (about $90 million) with progress payments, mostly from payload delivery services.

R&D partnerships and volunteer efforts are the most important sources of in-kind contributions. Previous sections already discussed the extent of the linkages that GLXP teams develop with R&D partners. Corporate partners can facilitate access to specialized facilities and resources such as workstations and expensive equipment. Intangible contributions comprise expert advice and hands-on experience with aerospace technologies. Academic partners can facilitate access to key equipment, laboratories and special facilities such as clean rooms or test areas for propulsion systems. Teams may also have that kind of access through the work of students and faculty members enrolled as team members. On the other hand the GLXP teams draw on serious volunteer effort in the form of direct labor (including students and friends for example) and sometimes in the form of production effort, parts and components. Team FREDNET for example estimates that that kind of volunteer effort may help to save up to $6 million in a $30 million mission (Kay, 2010). Another GLXP team refers to part of that volunteer effort as 'friend companies' and explains that it might ultimately represent between 5 and 20 per cent of the total mission costs.[55] Visits and direct observation of team workplaces suggest that the facilities accessed through family and friends are also very valuable to hold regular team meetings, meet potential sponsors or partners, assemble subsystems, or store equipment, parts and supplies.

Finding partners, sponsors and investors demands time and effort and diverts the teams' technology development efforts to great extent. This is particularly true for the teams that just entered the competition and in general for all teams when the competition is just launched. When the GLXP just launched it did not have much visibility and teams had to break an informational barrier and inform potential partners and investors not only about their projects but also about the competition. Moreover, the GLXP in particular was launched just before a significant global economic turmoil that discouraged all kinds of investment. Additionally newly entered teams generally do not have much to offer in exchange for support and may not have the appropriate business development skills to persuade potential partners or sponsors. Only after increasing visibility of both the competition and the teams ('after the team built its brand') are the teams able to raise interest and even choose the best partnership candidates or impose certain working conditions. Investors in particular are not willing to bear the risks of this kind of project up front, but only after there is some proof of technology feasibility and if high returns can be guaranteed.

There are still some factors that can facilitate partnerships and fund-raising. Team leaders generally point out that their teams gain credibility thanks to the brand image of the prize sponsor Google Inc. In principle potential partners and investors are generally pessimistic about the prospects of a project of this kind led by a small independent team despite the team members' experience and background. But the fact that a team successfully entered a competition sponsored by a company like Google Inc. and the involvement of this company with what is considered to be a leading edge activity helps more people to believe that the team's initiative is credible and that this whole enterprise makes sense. Working technologies developed by the teams also demonstrate the seriousness of the efforts and increase the interest of those willing to acquire those technologies. The GLXP also offers other valuable opportunities. Partners and sponsors can contribute technology to a challenging project and build heritage for their subsystems and components, something very important in space industry and other industries that produce systems and components for extreme environments. There is also a special marketing value for sponsors as the competition and the teams increase their visibility. This leads to more publicity and exposure to potential customers.

6.5.5   Constraints and Challenges

The GLXP teams face a number of constraints or specific challenges in the pursuit of their projects but they tend to minimize their importance. Limited time and the lack of resources to undertake their projects, both interrelated, are the most significant constraints that the GLXP teams face. All but two teams consider the limited time allowed by the prize a moderate constraint at least and about 40 per cent of them consider it a great constraint. The teams use (or plan to use) different strategies to overcome the limited time available to accomplish the mission and only one-third of them have thought to some extent of withdrawing from the competition because of this constraint.[56] Sixty per cent of the teams rely to a great extent on a strategy that involves the design of simpler technologies to speed up their projects. Another 40 per cent overcomes this constraint to great extent by using more existing technologies. A slightly smaller proportion seeks partners and only one-third of the teams increase their fundraising efforts. Skipping risk analysis/test phases is only considered to some extent by 40 per cent of the teams. To speed up their projects unconventional teams are much more likely to design new, simpler technologies than conventional teams. While this is the main strategy for three-fourths of the unconventional teams, only about 40 per cent of the conventional teams do that to a great extent. Conventional teams in turn are slightly more likely to use existing technology (about 40 per cent of them do this to a great extent) and partner with other organizations (30 per cent of them do this to a great extent) to speed up their projects. More generally team leaders speak of the importance of reducing engineering efforts to shorten achievement times and explain that a number of technologies needed for their projects are readily available at affordable price tags. An experienced engineer also explained that there is a limit to making things cheaper and the main approach its team uses to speed up developments includes drawing upon more volunteers and combining things to minimize the number of items the team builds (e.g. develop a model that serves both testing and promotional purposes).

Funding on the other hand can help technology development (and easily help to accomplish this kind of project within the given lead time, some team leaders say) but is generally not available to teams up-front. About 40 per cent of the teams consider this to be a great constraint and 30 per cent consider this a moderate constraint. Team leaders suggest that funding is more important than having the most advanced technology because without up-front funding there is not much that can be accomplished in this kind of project. Funding can buy not only existing technology but also expertise and solutions that otherwise teams would have to develop internally. The teams use (or plan to use) different strategies to overcome the lack of up-front funding but they rarely think of abandoning the competition because of this. At least 40 per cent of the teams design simpler technologies, increase their fundraising efforts and design marketable technologies to a great extent to overcome this constraint. The rest of the teams consider these options as well, but only to some extent. About half of the teams rely more upon existing technologies, skip risk analysis/ test phases or partner with other organizations but only to some extent. Interestingly unconventional teams also face the lack of funding as a constraint but are less concerned about it than conventional teams. More than 60 per cent of the unconventional teams also indicate that, when facing the lack of funding, they generally prefer to overcome the problem with the design of simpler technologies instead of using more existing technologies or seeking partners. On the other hand only 30 per cent of the conventional teams design simpler technologies to a great extent to overcome the lack of funding. They are actually more likely to use existing technologies but still not significantly. A third of the conventional teams tend to partner with other organizations when facing this constraint.

The next constraint in overall order of importance has been some ambiguous or questionable rules and technical requirements established by the GLXP, although opinions vary across teams. While about 70 per cent of unconventional teams do not consider this to be a constraint at all, more than half of the conventional teams are concerned about it. In particular the GLXP's MTA, which includes rules and other legal provisions for the competition, is the document that raised some controversy among teams. The initial version of this document (MTA 1.0) was drafted by the XPF and made publicly available to obtain feedback from the teams and the general public. In general the document leveled the field for competition but, from the point of view of the teams, it had several problems such as conflicting rules, rules that impeded free negotiation between teams and their partners, and difficult interpretation of technical requirements. This first set of rules also restricted potential innovations when they, for example, referred to wheeled rover-like capabilities and required the teams to take pictures of the rover's tracks on their 500 meter traverse (assuming that teams would use wheeled rovers and not other less conventional approaches). The next version (MTA 2.0) was introduced in August 2010 to obtain further feedback from teams. This time many issues were corrected or reformulated. In particular, issues with the interpretation of copyrights and technical requirements were fixed, leaving more room for teams to decide their approaches to the challenge. The latest major improvement (MTA 3.0 of January 2011) contemplated further feedback from teams to become 'more reasonable' and 'positive'. While teams generally consider this latest set of rules to be conducive, four teams still consider this a great constraint. Changes after the competition was launched caused discontent among teams, although in some cases there have been divided opinions. For example, the extension of the prize deadline from 2012 to 2015 was something positive for some teams (Werner, 2010), but other teams interviewed for this investigation (possibly those that progressed the farthest by that time) considered this to be disappointing.[57] More generally rules that change during the competition have negative consequences for the strategies of the teams. Two years after the prize announcement a team leader explains: 'I think a frustration for all the teams is the fact that the rules have not been finalized yet. I think that limits our ability to plan finally and decisively with regard to approach' (Goldsmith, 2009). Other changes that are related with the spirit of the competition or its ultimate purpose can also cause frustrations and disagreements between prize sponsors and entrants. In the GLXP at least one team withdrew from the competition during its initial development phases after an emerging disagreement between the team and the XPF on how the outcomes of this competition ought to be used and publicized.

ITAR regulations can become a significant constraint as they have two important implications for the GLXP projects. First this regulation impedes foreign teams sourcing a number of key technologies from USA companies such as propulsion, communications and navigation and control subsystems.[58] Foreign teams without access to certain technologies depend upon in-house capabilities or alternative non-USA sources if they are available. Second, this regulation penalizes the business plans of USA teams that seek to export their technologies. Multi-national teams are also affected, particularly when they adopt open-source approaches and seek to engage both USA and international members. More generally there are some sensitive technologies that are only available to USA government-funded projects and are not available to any GLXP team (as the GLXP mission must be privately funded).

Finally there are a number of other potential constraints that are generally not considered a concern by teams. The lack of knowledge or skills to pursue the project and the time advantage that first-to-enter teams have, interestingly are more a concern of conventional than unconventional teams. While almost 60 per cent of the former consider those to be somewhat important constraints, about 70 per cent of the unconventional teams say that those are not a constraint at all. Other potential constraints such as the competitive strategies of other teams or limited access to professional networks are not significant for most of the teams.

6.6   TECHNOLOGY OUTPUTS

Prize entrants may produce diverse types of technology outputs in their attempts to win the competition and in the pursuit of other goals. This investigation anticipated that the prize technologies can take the form of, for example, new concepts/designs, models or mockups, prototypes and actual spacecraft, which may or may not be commercialized during the competitions. The number, quality and degree of advancement that prize outputs represent are likely to vary across teams as they have different abilities and access to resources and have been involved in the competition for different periods. Prize teams may even begin their work in 'stealth mode' before officially entering the competition, that is, without making public their intention to compete. They may also have previous work experience as a group and even ongoing prize-related projects before even considering entering the prize.

The GLXP teams that participated in this investigation (17 teams) had a combined total of 450 months of technology development on their GLXP projects at the moment this analysis started. These teams entered

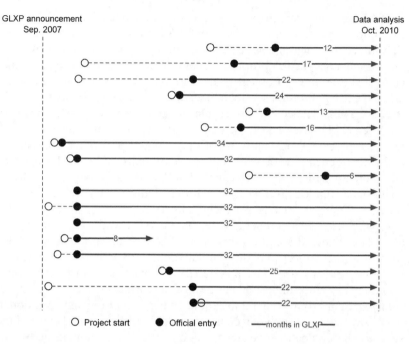

Note: N=17 cases.

Source: Questionnaire to GLXP teams.

Figure 6.5   Time in competition and time of actual development for GLXP teams

the competition between September 2007 and April 2010 but most of them started working on their GLXP projects months before their official entry. These teams have on average started their work about four months before their official announcement as GLXP teams. Two teams had even worked about 12 months before being announced as official competitors. More than half of the teams already had more than 30 months of work on the GLXP project when they responded to a questionnaire but none of them had yet completed more than 50 per cent of their GLXP projects (Figure 6.5). According to each team's plans to accomplish the mission, about 40 per cent of them completed only between 20 per cent and 50 per cent. The rest of the teams completed even less by that time.

The time the teams spend as official participants in the prize is not a good indicator to understand prize technology outputs for at least three reasons. First, there is an initial time period in the GLXP projects that teams use to hire members and gather initial resources to bootstrap their projects, and this set up period varies considerably across teams. Second, there are a number of teams that describe their GLXP projects as the continuation of past projects of the team or its members. About 40 per cent of the teams, mostly conventional, have past projects that relate to a very great extent to the GLXP and another 25 per cent have past projects that relate to the GLXP only to some extent. The rest of the teams, mostly unconventional, report that the GLXP is not related with previous projects of the team or its members (which is not surprising because these teams were not involved with the prize technologies before). When there is a relationship with previous projects, about one-third of the teams indicate that the GLXP is an expansion of ongoing projects and about 25 per cent of the teams indicate that involves the application of results or knowledge acquired in previous projects. One team indicates that the GLXP is the restart of discontinued projects. Third, there is a wide range of planned development lead times to complete the GLXP missions. The teams expect to complete their projects in time periods that range between 31 and 91 months with a median of about 57 months. There are not significant differences between unconventional and conventional teams in this regard.[59]

A survey of the technologies developed by 26 teams that participated in the first three years of competition shows that at least 16 of them produced some significant outputs, that is, systems/subsystems that are functional (but not necessarily flight-ready or aimed at mission accomplishment) and have been developed during the participation of the team in the prize. These technologies can help the teams to make different degrees of progress in the achievement of the GLXP mission. Ten teams have not introduced any significant technology output and four of them have already withdrawn from the competition. Some of those significant technologies might not have been developed if the GLXP did not exist. While almost half of the teams that participated in this investigation reported some probability or certainty about pursuing robotic planetary exploration projects regardless of the announcement of the GLXP, a similar proportion of teams indicate that they would not have worked in this area. Unsurprisingly conventional teams are those more likely to have pursued GLXP-like projects anyway. Slightly more than half of the unconventional teams are working on this type of project only because of the announcement of the GLXP. Only four out of nine teams exclusively motivated by prize incentives contributed some significant technology output and another two of them withdrew from the competition.

The overall examination of an illustrative (but not comprehensive) sample of the technologies developed in the first three years of competition shows their diversity (Table 6.7).[60] While some of these examples represent current-day technologies, most of them incorporate some kind of

Table 6.7　Selected examples of GLXP technology outputs in first three years of competition

Technology output (one row per team)	Lead time (months)[a]	Novelty[b]	Target[c]	TRL equivalent[d]
Outputs from unconventional teams				
No evidence of significant outputs	31	–	–	–
Stereo vision camera for 3D mapping and navigation	40	I	U	M
No evidence of significant outputs (team withdrawn)	31	–	–	–
No evidence of significant outputs	23	–	–	–
Solid-steam rocket motor prototype	25	N	E	L
Five-inch ball robot that can climb slopes	34	N	D	L
New, optimized software algorithms for systems error control and detection	24	N	C/H	M
Re-development of modular rocket system with parallel stages (25+ years old concept)	22	I	U	M
New air-launched, three stages orbital rocket	37	N	U	M
Lightweight, sealed and scalable N wheeled-motor design	29	I	U	M
No evidence of significant outputs	30	–	–	–
Ion motor-powered Cubesat	40	I	D	M
No evidence of significant outputs (team withdrawn)	37	–	–	–
Outputs from conventional teams				
Hopping moon lander concept (based on spacecraft previously built by partner)	35	N	C/S	M
Propulsion systems for new launcher, based on technologies previously developed by team members	3	I	C/S	M
No evidence of significant outputs	2	–	–	–

Table 6.7 (continued)

Technology output (one row per team)	Lead time (months)[a]	Novelty[b]	Target[c]	TRL equivalent[d]
Outputs from conventional teams				
Spinning lander concept (concept previously developed by team members) (team withdrawn)	8	N	U	M
No evidence of significant outputs (team withdrawn)	34	–	–	–
Leg-enabled robotic system prototype	34	I	D	M
Lander's rocket motor and navigation control for engine, based on COTS technologies	34	I	C/H	H
Software for autonomous rover's travel	19	C	U	H
No evidence of significant outputs	19	–	–	–
No evidence of significant outputs (team withdrawn)	37	–	–	–
Adaptation of NASA's Common Spacecraft Bus vehicle into lunar lander for payload delivery	39	C	C/S	M
Cubesats for demonstration of navigation techniques	37	C	D	H

Notes:
Selected outputs with illustrative purposes; based on data available to the author as of December 2010.
a. Months since project started (or since official entry if other data are not available)
b. C=Current-day technology, I=Significant improvements, N=New-to-industry technology
c. E=Only experimental (experimental technologies not necessarily used for GLXP project), D=Tech. demonstration/test for GLXP mission, U=Use in GLXP mission, C/H=Expect commercialization of hardware, C/S=Expect commercialization of services.
d. L=low (TRLs 1–4), M=medium (TRLs 5–6), H=high (TRLs 7–9)

Sources: Interviews with teams, GLXP website, team websites and press releases.

improvement or are new-to-industry (not conceptually, but in their actual implementation). This range of outputs includes for instance new propulsion systems, mobility mechanisms, demonstration Cubesats, camera/video systems and software. Most of these technologies are at medium and high maturity levels, the latter being closer to flight-ready. They are however aimed at different purposes. Within this set of 16 technologies six have been developed for direct use in the GLXP mission, four are aimed at technology demonstration or tests for the GLXP, three are aimed at commercialization of services, two are aimed at hardware commercialization and another one has been developed with experimental purposes. It should be considered that this analysis is however a somewhat limited description of the entire range of GLXP technology outputs. In addition to these examples there are many more significant intermediate outputs, incremental improvements and ongoing developments that might ultimately result in new components, parts or systems for the GLXP mission or other projects. Those kinds of prize technologies are more difficult to track without a more in-depth survey that was out of the scope of this investigation. There is still evidence of their existence. As previously described some teams adopt R&D approaches that draw more heavily on, for example, trial and error or technology adaptation, which helps to solve specific technical problems, produce useful mission data or adapt commercially available components. All these activity and their outputs are contributed by not only the teams but also their partners and collaborators.

The evidence also shows some variation in terms of kinds and number of technology outputs across teams. A few unconventional teams for example have introduced lower TRL technologies for experimentation and demonstration purposes, but it is not clear whether these technologies will ultimately be deployed or commercialized. For example, there are teams such as Selene that are exploring concepts to find the simplest, most reliable and cost-effective method to travel the required 500 meters, including the 'classic planetary rover design' (GLXP, 2011a). On the other hand the conventional teams have been more likely to produce technologies aimed at delivering launch, payload and/or other services, that is, at higher TRL. Several of those technologies were already under development prior to the GLXP by the team or a partner, are under development at NASA and will be adapted for the GLXP (through for example Space Act Agreement partnerships) or are based on proven designs of past space programs.

A few illustrative examples of technologies developed by GLXP entrants help to better demonstrate what kind of outputs the prize can induce (Figure 6.6). The first example shows the rover prototypes Asimov Jr. R1 and Asimov Jr. R2 introduced in 2009 and 2010, respectively, by the team Part-Time Scientists as a result of different iterations to find the most efficient design to accomplish this mission (Figure 6.6a).[61] The second example shows the prototype Red Rover developed by team Astrobotic which features, among others, a newly developed and advantageous

(a) Prototypes Asimov Jr. R1 and Asimov Jr. R2 by team Part-Time Scientists

(b) Prototype Red Rover by team Astrobotic

Source: Part-Time Scientists/ Alex Adler.

Source: Astrobotic Technology Inc.

(c) Illustration of team ARCA's IAR-111 Supersonic Carrier Plane

(d) Snapshot of 3D simulation with team Selenokhod's lunar rover

Source: Romanian Association for Cosmonautics and Aeronautics (ARCA).

Source: Selenokhod/ SmirnovDesign.

Figure 6.6   Selected examples of GLXP technology outputs

internal motor-mobility mechanism (Figure 6.6b).[62] The third example is an illustration of team ARCA's multi-purpose IAR-111 Supersonic Carrier Plane for the Haas II rocket (also being built by the team) that the team plans to use to launch its GLXP mission (Figure 6.6c).[63] The last example shows a snapshot of a 3D simulation with team Selenokhod's lunar rover, a two-skied solar powered craft under conceptual development that might, thanks to its novel design, overcome typical problems that other types of mechanisms face in extreme planetary environments (Figure 6.6d).

The wide network of partners and collaborators examined in previous sections already anticipated that many of the GLXP technologies would actually be produced by team partners or in collaboration with other organizations, particularly in the case of conventional teams. A handful of examples also show that technologies may be already under development or produced specifically for the prize. Team Odyssey Moon is developing its lander based on the engineering and technical expertise provided by NASA's Ames Research Center under the development of the Common Spacecraft Bus lander program. Team Rocket City Space Pioneers is developing its own propulsion systems based on the work of a company previously owned by the team leader.[64] These systems may eventually be used in the team's GLXP mission, but the team is also aiming at creating a sustainable space business (GLXP, 2010a).[65] The hopping lander of team Next Giant Leap is under development at Draper Labs with laboratory resources and labor that includes students from the Massachusetts Institute of Technology (GLXP, 2011b). The evidence also suggests that innumerable other components and parts included in the systems developed by the GLXP teams are outsourced. Team White Label Space for example is drawing upon the work of Lunar Numbat, one of its partners, to develop an engine throttle controller as an open source effort (White Label Space, 2010).

Although the evidence shows significant technology outputs induced by the prize, only one-fourth of the systems designed for the GLXP are completely new according to the teams, that is, they are built by the teams from scratch rather than acquiring, adapting or copying from existing technologies. Overall about 45 per cent of the teams describe their technologies as somewhat new and 30 per cent as not new at all. In particular lunar landers and rovers are the most novel developments (i.e. completely new) as 50 per cent and 45 per cent of the teams indicate, respectively. The systems that follow in terms of novelty are the photo/video system, control/navigation hardware and software and Earth-to-Moon transfer vehicles. Twenty-five per cent of the teams indicate that these three subsystems are completely new. Another 45 per cent of the teams indicate that their Earth-to-Moon transfer vehicles and the ground support systems are not new at all. Earth-Moon communication systems have a similar degree of novelty. Notably unconventional teams are more likely to use infrastructure (i.e. communications and ground support systems) and control/navigation hardware and software that are not new at all, as more than half of those teams indicate. Conversely conventional teams are more likely to develop lunar rovers that are completely new and use Earth-to-Moon transfer vehicles that are not new at all, as 70 per cent of them indicate in each case.

A relatively low degree of newness in the overall outputs of this prize

Table 6.8   *Number of GLXP teams that achieved significant innovations, by type of entrant*

Type of innovation	Number and type of team with innovations, by form of achievement				Total teams
	Unconventional		Conventional		
	Planned	By chance	Planned	By chance	
New products	5	1	3	1	10
New ways to organize tech. design and dev.	3	2	–	1	6
New use for existing materials, products or components	3	2	1	2	8
Other innovations	1	–	–	–	1
Total teams	5	2	3	4	10

Note: N=15 cases; cells indicate number of teams that achieved significant innovations of each type, based on self-reported data.

Source: Questionnaire to GLXP teams.

does not mean that innovations do not occur. At least 10 teams consider that they have achieved significant innovations in the form of new products or components (ten teams), new uses for existing materials, products or components (eight teams) or new ways to organize technology design and development (six teams) (Table 6.8). These innovations have been achieved either purposely (eight teams) or unexpectedly (six teams). Interestingly all the teams that reported significant innovations also indicated that their innovations are useful not only to accomplish the GLXP mission but also to pursue other projects. Unconventional teams report the achievement of significant innovations more often than other teams, innovate mostly in terms of new products and plan most of their innovations. The high proportion of innovating teams suggests a word of caution in the interpretation of the data. Some of these innovations might actually be new designs that the teams are exploring and not actual technologies that the teams are introducing. Interview data confirmed that this is the case with a few teams, but, in most cases, innovations refer to new-to-industry or 'new-to-industry in this world's region' (i.e. Europe) subsystems that teams are implementing in their own missions.[66]

The examination of technology outputs by type of entrant suggests emphasis on different points of the innovation pathway (Figure 6.7a and Figure 6.7b). While the developments of unconventional teams tend to

a) Unconventional entrants

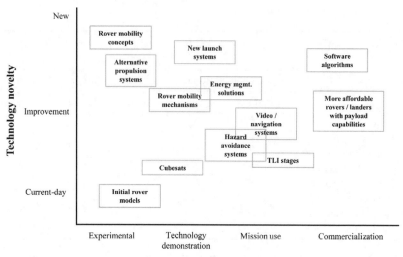

b) Conventional entrants

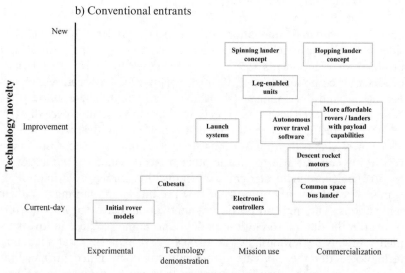

Source: Author's analysis.

Figure 6.7   Selected technology outputs from first three years of GLXP in novelty/maturity scales

gather around the area formed by improvements/new technologies and experimental/mission uses, several technologies developed by conventional teams are at higher TRL, closer to potential commercialization in the form of service delivery or, in a handful of cases, hardware sales. Only a few of all these technology outputs might become true breakthroughs for aerospace application (e.g. new lander concepts) or broader industrial use (e.g. software algorithms). Other technologies, particularly those aimed at providing services (e.g. payload delivery), are based on improvements of current-day technologies or technologies already under development. Several significant improvements in terms of mobility mechanisms and electronics/navigation systems will be used in GLXP missions. Some unconventional teams are also testing multiple conceptual designs of mobility systems (e.g. rovers or other units) and alternative propulsion schemes (e.g. thrusters, fuels).

Although the evidence shows that a majority of the prize technologies are not aimed at or readily available for commercialization/service delivery, we shall consider that most of the GLXP teams seek to generate revenues to fund their projects. Some of them have been successful in this regard. On October 2010 for example NASA awarded up to $30 million in ILDD contracts to six USA teams to purchase technical data resulting from the development and demonstration of capabilities of robotic lunar missions (NASA, 2010d).[67] The awardees include team Astrobotic, team Next Giant Leap (through its partner The Charles Stark Draper Laboratory Inc.), team Rocket City Space Pioneers (through its parent company Dynetics Inc.), team Omega Envoy, team Moon Express and team FREDNET. Commercialization strategies also contemplate revenue models based on products other than hardware sales, payload delivery or IP licensing. About a year after the prize announcement, for example, team Odyssey Moon reported that it sold 75 per cent of the available payload capacity of its spacecraft to five customers (Odyssey Moon, 2008). In their effort to raise funding or profit from their projects, teams also commercialize/plan to commercialize sponsorship opportunities, expertise/project management services and a variety of final consumer services related with social networking, photography and video and remote exploration experiences. The following are examples from team interviews, team websites and press releases:

1. *Sponsorships*
   Includes for example the commercialization of the naming rights for a team's space vehicles, mission or other appearances of the team or its members. It may also include a shared team logo (to the extent allowed by GLXP rules).

2. *Communications*
   Chat, voice, e-mail and social networking via/with team's spacecraft.
3. *Payload delivery*
   Payload delivery with scientific (e.g. scientific instruments), commercial (e.g. human remains, personal objects, corporate-related instruments) or artistic (e.g. objects, inflatable sculptures) purposes.
4. *Team expertise*
   Consulting or project management services based on expertise from GLXP mission.
5. *Moon experience*
   Consumer-oriented services such as tele-presence through 3D video and images, remote driving of rovers.
6. *Hardware*
   Commercialization of technology through agency or corporate contracts.

These diverse revenue sources have different strategic implications for mission financing. Data licensing and Moon experience commercialization for example require the mission to be executed to generate revenues. Payload delivery services and sponsorship opportunities can generate revenues through payments in advance (this is typical of the space payload delivery business). The value of expertise/project management services is likely to be associated with the team's performance in the competition. Other sources of revenue are linked to spacecraft's technical capabilities. The commercialization of Moon experiences through 3D video and remote driving of the rover, for example, requires longer mission times and technical capabilities to perform certain functions not necessarily required to claim the prize reward. This investigation cannot foresee all the alternative schemes that the teams will pursue to commercialize technologies after the prize has been won, but it is possible to anticipate a wide range of approaches. These approaches may range from direct commercialization of hardware through licenses to the creation of consortiums of companies to produce and commercialize the technologies on the basis of co-owned IP.

A key factor to explain prize outputs is also the lack of a requirement to place the technologies developed by entrants in the public domain, which enables not only the R&D activities of the teams but also the commercialization of their technologies. The rules of the prize still allow the XPF to retain media rights on certain GLXP developments and the eventual mooncasts of the winning missions.

There are diverse sources and uses of IP rights in the GLXP.[68] The teams that have access to university resources such as laboratories and machine shops enter/plan to enter in agreements to license the technologies developed at universities. Some teams have created more dense networks of corporate partners to source key technologies in exchange for IP on other technologies they have developed. The GLXP teams that have signed ILDD contracts for the provision of mission data give away part of their IP to NASA. The teams that are organized as workgroups, use collaborators and volunteers and have other types of partners enter in agreements to share the IP on parts and components developed for the prize. A few team leaders explained in interviews that they will eventually seek to file patents for some of their technologies, but the cost of doing that has been a constraint. The teams generally define themselves as 'fairly open' with regard to knowledge sharing. Most of them publish progress updates on their websites and the official GLXP website. The teams are more likely to disseminate knowledge on methodologies, technical approaches to solve specific problems and test results, but sometimes they also disclose or openly share key technical details and other project management information. An overall assessment of the digital media published on the Internet (e.g. short videos on progress updates) suggests that unconventional entrants might be more inclined to share this kind of detail, particularly when their goals are not related with the creation of a commercial enterprise.

NOTES

1. The GLXP website originally hosted an online forum open to the public as well. According to the XPF, 'the small team that works on GLXP simply ran out of the time resources required to moderate such a large and active forum community' (GLXP, 2009). Therefore the forum was moved to a third-party, space specialized website (SpaceFellowship.com GLXP Forum).
2. The GLXP's rules provide that if a government-funded mission launched after 1 January 2010 performs a similar mission, the Grand Prize will be reduced to $15 million. Space agencies from countries like India and China already have ongoing programs that may include Moon surface exploration. There are also new NASA initiatives that seek to send robotic spacecraft to the Moon. NASA's 2011 budget proposal, for example, considers lunar expeditions that would test the ability to control robots remotely from Earth or the International Space Station and transmit near-live video (Hsu, 2010; NASA, 2010b; Werner, 2010).
3. The actual amount paid for each Bonus Prize will depend on whether other Bonus Prizes have already been paid (the total will be $4 million).
4. Teams are allowed to use other government resources yet those resources will be considered public funding and will count against the maximum 10 per cent for public financing.
5. More information about this competition is available from California Space Education and Workforce Institute (2009).
6. More complex or 'tightly coupled' designs increase the probability of facing technical problems because there are more parts and components and more complex

interrelations between them, and the probability of detecting a problem before suffering significant damages is much smaller.

7. The Mars Sample Return NASA/ESA program, aimed at collecting samples from Mars and returning them to Earth, is an illustrative example of the development cycle-cost relationships. The preliminary planning for this mission began in 2008 and the mission actual launch is expected by 2018–22 (that is, more than 10 years of development). The total expected cost of its multi-element, more complex mission is about $7 billion (iMARS, 2008; NASA, 2010c).

8. SpaceX is one of the most illustrative examples. Since 2002 and during its first six years (which the company used to develop its first commercial launch rocket from scratch) this company has been backed almost entirely by NASA and its founder's own money (Homans, 2010). Another interesting example is Armadillo Aerospace, the winner of the NGLLC. This startup was awarded $475 000 to perform test flights of its experimental vehicles under NASA's Commercial Reusable Suborbital Research Program (CRuSR) (NASA, 2010e).

9. The Delta Clipper DC-X is a technology demonstration project to develop vertical take-off and landing vehicles in the 1990s. The project was first led by the U.S. Department of Defense and then by NASA, with participation of McDonnell Douglas Corporation.

10. Recently launched, new US companies started by executives from non-aerospace businesses over the last 15 years comprise for example Blue Origin, XCOR, Armadillo Aerospace, Masten Space Systems, and SpaceX.

11. Small missions have been the mainstream concept for the new space sector since the first private, commercial, and student-oriented initiatives of the 1990s (Ridenoure and Polk, 1999). Space agencies and large corporations have increasingly adopted this approach. Since the 1990s NASA has introduced a series of small, fast-track implementation Discovery class missions at low (and sometimes capped) cost. The NEAR Near Earth Asteroid mission and the Mars Pathfinder mission are examples of that kind of approach. Large companies such as Hughes Communications Inc. and Rockwell International also worked on small landers development during the 1990s for Mars exploration programs (Spear, 1995; Vorder Bruegge, 1995). There is also an entire small satellite industry based on this approach to space exploration (Bonin, 2009).

12. Unless otherwise indicated, the analysis is based on the following scholarly articles and reports: Zakrajsek et al. (2005), Balint et al. (2008), NRC (2008), Braun (2010), Parabolic Arc (2010), Seeni et al. (2010), and XPF (2008b).

13. The teams might also launch their spacecraft as a secondary payload to share costs with other missions, but providers may be unwilling to accept risky payloads that have their own propulsion systems for Earth-to-Moon transfer (Werner, 2010). The XPF also considered offering a smaller cash purse when designing the prize, and providing a launch vehicle for the first few teams to be launch-ready, but the idea was discarded to offer a more open-ended definition of the challenge and promote innovations in mission approaches, among other reasons (XPF, 2011a). The AXP already demonstrated that open-ended definitions of the prize challenge may lead to the introduction of unconventional approaches. Instead of developing a conventional rocket, the GLXP teams might try unorthodox launch methods such as using balloons or spaceplanes with horizontal take-off.

14. SpaceX is one of the top players in this market and is a preferred launch partner of the XPF. The company offers discounts of up to 10 per cent in its Falcon 1e and Falcon 9 rockets to GLXP teams (SpaceX, 2011a, 2011b). It should be noted that there has been a market of launch vehicles with similar capabilities and relatively equivalent costs since at least the 1990s (Poniatowski and Osmolovsky, 1995).

15. For the reader's reference, NASA's Mars Exploration Rovers were designed to drive up to 40 meters per Earth-day, and that goal was notably exceeded (NASA, 2009b).

16. For the reader's reference, sending one gigabit (Gbit) of data from the Moon to the

Earth may take up to 24 hours. That amount of data is equivalent to the transmission of about 100 pictures taken with a regular point-and-shoot digital camera.

17. Rovers may operate longer than originally planned and then generate further operational costs that are not considered here.

18. This technology was used in, for instance, the 1970s Soviet Lunokhod Moon rovers to hibernate through many nights, keeping the craft's interior sealed and warm.

19. NASA's Technical Reports Server (NTRS) is available at: http://ntrs.nasa.gov/search. jsp. To illustrate the kind of knowledge available on this database, a quick search of the keyword 'lander' matches 9362 documents.

20. The analysis of technology scenarios for mission approaches (programmatic-level) is based on Pedersen et al. (2002), Fink et al. (2005), Seeni et al. (2010), and NASA's documentation on past missions as cited in this section. System-level analysis of technology scenarios is based on NASA's technology roadmaps (NASA, 2011).

21. Peter Diamandis, founder of the XPF, was CEO of this company.

22. Interestingly part of LunaCorp's team has entered the GLXP with Team Astrobotic. Both Mr. David Gump and Dr. William 'Red' Whittaker from Carnegie Mellon University were part of the teams of these two projects, the former as the president of the company and the latter as team leader.

23. Full-time members spend 80 per cent or more of their work time in the GLXP. Part-time members share most of their work time with another job or activity.

24. Team FREDNET decided not to participate in this study. There is another GLXP team that pursues an open-source approach but its membership is considerably smaller.

25. FREDNET is the team that has grown the most in that regard. The team enrolled about 20 people within the first week of participation in the GLXP and then grew to about 100 people in a few months (Evadot, 2009).

26. The total adds up to more than 100 per cent because questionnaires allowed respondents to indicate more than one type of educational background for each team member.

27. Another dimension of diversity in team membership is age. Team FREDNET is again an interesting example. This team reported eight- to 80 year-old volunteers that contribute remotely to the project (Evadot, 2009).

28. Extracted from interview with members of the GLXP team T16.

29. The author also contrasted this classification with other data available for each team (such as team websites and interviews) to verify that it properly reflects how familiar the teams are with the prize technologies. When the analysis required the classification of teams that did not participate in the study (for example, to assess technology outputs) the researcher analyzed other data that were publicly available about the team and type of entity (i.e. for-profit, non-profit or independent). Nine teams that entered before December 2010 did not participate in this study. Five of those nine teams are companies or groups thereof and have significant aerospace experience. These five teams are considered conventional. There are also three non-profit organizations or independent groups and one team organized as a group of companies and NGOs with no apparent space agency or industry experience. These four teams are considered unconventional.

30. Extracted from interview with members of the GLXP team T4.

31. Extracted from interview with the XPF's Director for Space Prizes (Pomerantz, 2010a).

32. This description of the work of engineers is based on an interview with members of the GLXP team T20.

33. Extracted from interview with the leader of the GLXP team T16.

34. Extracted from interview with members of the GLXP team T4.

35. Expressions extracted from interviews with the leaders of the GLXP teams T4 and T11.

36. Extracted from interview with the leader of the GLXP team T11.

37. The story told by a young engineer in interviews is illustrative. Very excited after attending a GLXP team's project presentation at his university, he decided to join the team and focus his doctoral dissertation work on the development of a technology that the team could use in the GLXP mission. Part of his family did not share his enthusiasm at the beginning because they considered the project unfeasible and undeserving of the

effort he was making. His participation was questioned, however, only until his family attended another presentation given by his team at an international aerospace exhibition. The quality and seriousness of the presentation, he explains, ultimately convinced his family about the value of the pursuit.

38. Extracted from interview with the leader of the GLXP team T20.
39. Extracted from interview with a member of the GLXP team T16.
40. Extracted from interview with the leader of the GLXP team T20.
41. Extracted from interview with the leader of the GLXP team T11.
42. An example of this is a team that has screened commercial-grade components in a cryogenic freezer to simulate extreme lunar temperatures and determine which components are capable of bouncing back from the extreme deep freeze. This kind of test allowed the team to identify batteries, solid state drives, and processors that are capable of resuming operation when the temperature warms back up to roughly minus 80 degrees Fahrenheit (Astrobotic Technology, 2010).
43. Extracted from interview with the leader of the GLXP team T16.
44. These are subsystems specifically mentioned by GLXP team leaders in interviews.
45. Extracted from interview with the leader of the GLXP team T4.
46. The author attended the 4th annual GLXP Summit and took note of discussions between GLXP team members to share the launcher and split its costs between teams.
47. A word of caution is necessary in the analysis of partnerships. The underlying data underestimate the real number of partners because (as team leaders confirmed) not all partnership agreements are publicly announced. Moreover, the researcher detected inconsistencies in the number of partners reported by alternative data sources that cannot be attributed to normal variations related with team growth.
48. Notably, the Charles Stark Draper Laboratory, a US non-profit R&D lab, is partner with two competing teams Next Giant Leap and Rocket City Space Pioneers.
49. That is the case of team Odyssey Moon, which signed a Reimbursable Space Act Agreement with the NASA Ames Research Center whereby NASA provides technical data and engineering support to the team to develop a lunar lander and the team reimburses the costs and shares the data from tests and actual lunar missions. In fact all GLXP teams can under a Space Act Agreement access NASA's engineering and technical expertise but subject to ITAR regulations (MacDonald and Marshall, 2008).
50. Range based on data gathered from five different teams. The estimate for other teams is very likely to be within the same range.
51. This analysis does not detail all the cost components of each phase, which may comprise labor, parts, third-party services, and insurance. Some of these components represent major expenditures. The XPF estimated for example that insurance costs for this kind of mission will be up to $2.5 million (XPF, 2008b).
52. For example, SpaceX, preferred launch partner of the XPF, offers discounts of up to 10 per cent in its Falcon 1e and Falcon 9 rockets, which cost about $10 million and $50 million before discount, respectively (SpaceX, 2011a, 2011b).
53. Extracted from interview with an engineer of the GLXP team T11.
54. The latter is the case of White Label Space. As part of its business strategy, the team chose a placeholder name that a sponsor will eventually replace with its own brand.
55. Extracted from interview with a member of the GLXP team T4.
56. This investigation probed in particular seven potential responses of teams to overcome constraints, namely: design of simplified new technologies, increasing use of existing technologies, increasing fundraising efforts, pursuit of more partnerships, design of technologies that can be commercialized, skip risk analysis/test phases, and withdraw from the competition. In interviews and document analyses the author sought additional data and explored other strategies of the teams to overcome constraints. Those strategies are all described in the text.
57. The change of the prize deadline (from the original 31 December 2012) was among the major changes in this process of crafting the rules. The XPF communicated that the deadline was extended because the process of finalizing the prize rules took longer than

expected and the economic conditions had not helped the participating teams to raise funding for their projects.

58. These are subsystems specifically mentioned by GLXP team leaders in interviews.

59. Lead times reported by conventional teams range between 31 and 91 months, and those reported by unconventional teams range between 38 and 81 months.

60. To examine the technology outputs of the GLXP this investigation draws on the information that the teams have made publicly available, which might result in the underestimation of actual outputs. The list with examples of technology outputs that this section shows does not seek to be comprehensive but illustrates the diversity of technologies that teams can contribute to accomplish the GLXP mission or other goals. The examples are selected from interview data and data collected from team websites by selecting at least three of the most widely publicized developments of each team. The selection of examples also seeks variation in types, use, and maturity level of technologies.

61. In late 2011 the team introduced a third iteration of this rover, Asimov Jr. R3. While the first two prototypes were built using rapid prototyping technology and 3D-printed FDM parts, the latest version is mostly built in aluminum.

62. The motors and mobility mechanisms, rather than being situated inside the rover wheels (as seen in other traditional designs) are inside the main body of the rover, which has advantages for surviving the extreme lunar environment.

63. This supersonic carrier is not only useful to launch the Haas II rocket but also to target the space tourism market.

64. The company previously owned by the team leader was acquired a few years ago by Dynetics, the leading corporate partner of this team formed by a handful of companies.

65. Team Rocket City Space Pioneers even suggested to other teams to share the launch vehicle.

66. Concern about the interpretation of the meaning of innovation in questionnaires was discarded when the author was able to discuss this with engineering-background team leaders and discovered that they acknowledge the difference between invention and the traditional concept of innovation (simply defined as creation and commercialization) and understand how that relates to the work they are doing.

67. The application process for ILDD contracts was also open to foreign teams and some of them actually submitted applications. None of them was awarded however.

68. The author was able to gather data on the IP strategies of teams only through interviews and team websites. Although not all teams were surveyed, those data still illustrate the diversity of strategies across teams and the effect of implementing a prize that allows entrants to retain intellectual property rights on their technologies.

7. Discussion

7.1 PRIZE INCENTIVES AND THE MOTIVATION OF ENTRANTS

To probe the ability of prizes to target and attract certain groups of interest by offering a particular set of incentives, this investigation put forward a hypothetical explanation of the decisions to participate in prizes based on distinct perceptions entrants have according to their experience with the prize technologies (H1). That is how, hypothetically, outsiders perceive opportunities to participate in projects they would not have access to and industry players come across a chance to exploit skills and technologies they are already familiar with. While there is no evidence to completely discard this explanation, the decisions to enter prize competitions and the ability of prizes to attract entrants are more complex phenomena. A key element in a stronger explanation of the incentive effect of prizes is their ability to capture the attention of a wide range of individuals and organizations with diverse goals and distinct perceptions regarding the opportunities offered by each competition. The most interesting of this is that those goals and perceptions are not necessarily associated in a direct manner with monetary values or the achievement of the prize challenge. The evidence that supports this explanation shows that, controlling for possible alternative strategies to achieve the prize challenge, there are entrants that make efforts that do not get them closer to the finish line or divert their attention from the prize ultimate goal. In the GLXP this can be observed for example in the indifference of some entrants concerning the strategies and advantage of other teams, projects that do not contemplate Moon landing or R&D activities to develop technologies with capabilities that exceed or do not exactly target the prize requirements. Furthermore, the entrants that fall into these exemplar situations can easily outnumber other entrants that are mainly guided by their desire to win the competition and therefore focus on the fastest possible achievement of the prize challenge. This disconnection between prize participation and R&D activities that are actually aimed at meeting the prize requirements has important implications not only for understanding the prize phenomenon but also for efficient prize design.

To better understand how prizes attract entrants let us consider first both the monetary and non-monetary incentives offered by prizes, which are remarkably diverse. In the GLXP the monetary incentives include the total $30 million in cash purses and other equivalent in-kind benefits such as access to discount price services from the XPF's preferred partners. Despite the hefty amount of prize money only a minor share of teams is primarily motivated by the cash purse or valuable in-kind benefits. These are generally the entrants that seek to create a commercial enterprise because, although the cash purse cannot cover the costs of a typical GLXP mission, a business case based on the prize technologies can reach sooner a break-even point if the team can get the prize money and use the prize in-kind contributions. Second place and bonus prizes then create a more sustainable incentive as runners-up can still win some money even if they were not the first to accomplish the challenge. This kind of distribution of prize money was in fact purposely set by the XPF to allow a second set of teams to continue attracting the interest of investors and engaging more members and volunteers (Pomerantz, 2010a). The AXP and NGLLC's total prize money (worth $10 million and $2 million respectively) had a similar effect on entrants that sought to create a company to serve markets related with the prize technologies. The cash purse of these three prizes may have also had the promotional effect suggested by the literature (for example Diamandis, 2009) and prize experts (for example Davidian, 2010) and may have helped to raise public and media awareness due to their significant amounts. The GLXP's monetary awards not linked to prize performance (i.e. Diversity Award) do not seem to have a significant effect on the decisions to enter the prize.

On the other hand entrants perceive diverse non-monetary incentives in these prizes. First, there is the opportunity to participate in a technically challenging project. This drives curiosity and creates the desire to compete. It also represents an opportunity for those who seek to learn and gain experience with technology development. Second, there are the reputation and publicity values created by the competition. Popular and highly regarded competitions such as the GLXP, thanks to the brand image of its sponsor, Google Inc., help to gather resources to accomplish the prize challenge or pursue other goals such as technology commercialization. The AXP also had the benefit of novelty and value of being possibly the first important modern prize competition inspired in this old incentive mechanism. Moreover, participation in these competitions has generally been an opportunity for entrants to build reputation, demonstrate technological leadership and promote their prize or other efforts. Third, these prizes also represent an opportunity to pursue other diverse personal and organizational goals. The GLXP for example has helped to organize other

activities of prize entrants, enabled the application of pre-existing technical concepts and entertained space enthusiasts.

The publicity/reputation value prizes create however is likely to vary
across technology fields. Prizes can make visible the activities and competitive performance of newly created startup teams so that they can be
followed by space agency and corporate officials. This also contributes
to building teams' reputation by highlighting their achievements and ultimately distinguishing the winner and runners-up from the rest of the participants and other non-participant organizations in the same sector.[1] In
these three prizes this exposure is associated with the promotional efforts
undertaken by the XPF and, more generally, the symbolic value of being
recognized as a competitor. In the GLXP it is also associated with the
value of the Google Inc. sponsorship. But particularly in these aerospace
prizes this phenomenon relates with the existence of high entry barriers to
a space sector traditionally dominated by space agencies and large companies with vast experience in aerospace development. The most interesting in these cases is that the low entry barriers to the competition make
this opportunity to gain reputation and publicize efforts widely available
to all entrants, including those that are not familiar with technology
development and those that serve other technology markets.

These prizes (particularly the GLXP and AXP) have been also associated with potentially sizable technology markets that are important for
entrants that seek to create a commercial enterprise based on the prize
technologies or other related markets. Over time however the perceptions
about the value of the technologies may change as the competition develops and some entrants successfully market (or not) their inventions. This
is particularly relevant in prizes with long entry periods such as the GLXP
in which late entries can learn and gather more information by observing
the activity of others that entered early in the prize timeline.

The examination of types of entrants and the relationship between them
and the incentives offered by the prizes help to better explain the ability of
prizes to attract outsiders and learn more about prize entrants in general.
In the GLXP unconventional entrants are generally organized as new,
non-profit or independent groups. They are on average larger teams and
tend to draw more on volunteer effort and students. Their most important
motivations to enter the prize are associated with opportunities created
exclusively by the prize such as the chance to participate in a challenging
project and learn. They are also very interested in the social benefits of
their projects. Some of these entrants consider the potential commercialization of technology but that is generally not their priority. The linkage
between prize incentives and outsiders is also given by evidence that
shows that more than half of the unconventional teams would not have

worked on this kind of project if the GLXP had not been announced. These teams consider that there are no significant constraints for their projects, the most important constraint being the lack of time to accomplish their space missions. They also tend to be less risk-averse and mostly concerned with an excessive financial exposure. The AXP and NGLLC also attracted mostly unconventional entrants. In these cases however the most prominent motivations were linked to the market value of the technologies and the prize money respectively. The AXP also represented a unique opportunity for Scaled Composites to publicize its efforts and demonstrate the capabilities of its technology to industry and the general public. The NGLLC was not as popular as the AXP but it also represented an opportunity for these generally unconventional teams to demonstrate their ability before the very sponsors of the competition that were potential customers of the prize technologies. The range of AXP entrants was as diverse as the GLXP's. There were significant volunteer efforts and also companies that re-directed their activities. The only evidence about their risk perceptions is related with the possibility of the prize money not being paid (which can be linked to the fact that this was the first prize offered by the XPF). In the NGLLC the unconventional entrants were very small teams that draw generally on their own resources. With a few exceptions these were newly created teams or pre-existing teams that redirected their efforts. These teams were more concerned with the risks of economic loss, excessive time spent on the project or the possibility of not being paid the prize money.

On the other hand the GLXP conventional entrants are generally organized as for-profit teams. At least half of them existed before the prize announcement. On average they are smaller and draw less on volunteer effort and by definition have more experience with the prize technologies. Their participation can be associated not only with the perception of the market value of the prize technologies but most importantly with the perception of an opportunity to gain recognition from NASA and other space agencies for potential future contracts. This opportunity is associated with technology commercialization but is also by definition an incentive created by the prize. These teams also consider using the prize technologies in other projects. The cash purse is important for some of these teams as well. They generally perceive more significant constraints such as the lack of time or up-front funding, unclear or ambiguous competition rules and the time advantage that other teams have. They are also more risk-averse and particularly concerned with financial risks and diverting efforts from other team activities. In the AXP there were also a few teams with space agency/industry experience and diverse organizational forms. Their ultimate goal was generally the creation of a commercial enterprise. They entered the prize either to gain reputation to attract investors or pursue the prize money as a source of funding. Only one of these teams drew on significant volunteer effort. They considered the lack of up-front funding the most significant constraint for their projects. The NGLLC did not attract entrants with traditional space agency/industry background (the two winners can be considered emerging space start-ups).

There are four key underlying factors that can generally explain why entrants with such diverse goals and perceptions decide to participate in prizes and choose these three prizes in particular. The first two factors have to do with entrant-level attributes and the other two are generally associated with the prize phenomenon.

1. *Shared beliefs*

    Each prize encompasses beliefs that prize organizers and entrants share about both the technical feasibility of the project and the merit of the pursuit from the personal, social or market perspectives. The source of these beliefs is with the entrants and their more general group or individual perceptions, based on their own knowledge, previous experience or even ingenuity. If entry barriers are low, prizes may even attract individuals and organizations that do not have experience with the prize technologies or related skills. This is a key factor in the explanation of the ability of prizes to attract outsiders. In the case of space prizes for example there are more space exploration enthusiasts than professionals employed in the space industry. Experts have also suggested that this might be part of a broader social phenomenon related with the existence of a new generation of entrepreneurs with non-aerospace backgrounds inspired by the space exploration possibilities that were not fulfilled in the past, a phenomenon that might not have an equivalent in other sectors.

2. *Lower risk aversion and optimism*

    Individuals and organizations that enter prizes tend to be less risk-averse than established industry players, but they can also be overly optimistic in their predictions about the chances of winning the competition, the market prospects and particularly about the commercial viability of their projects. Despite the uncertainty that experts perceive in the characteristics and value of the markets associated with these competitions, to mention only an example, there still have been entrants that seek to create a commercial enterprise based on the prize technologies.[2] Positive trends in the space sector can also reinforce entrants' perceptions. The new space industry and movements toward a more open space activity for example offer inspiring success stories and demonstrations that the pursuit of low-cost, low-technology space projects is possible. But there are other factors involved in this. Team members are generally self-motivated and proactive individuals that enjoy being involved in a challenging project which can explain at least in part why they participate. The evidence that suggests lower risk aversion and optimism can be also interpreted as both the existence of goals beyond the prize and significant benefits of prize participation regardless of performance. This can explain why for example most of the GLXP's unconventional teams, unlike the conventional ones, consider that there are no risks involved in prize participation and that their lack of knowledge and skills or the time advantage that first-to-enter teams have are not a concern at all. Most of these teams are motivated by reasons other than the prize money and, regardless of their competitive position, still benefit from the opportunity to expand their activities, professional relationships and personal experiences. Conversely conventional entrants already work on related projects and consider the competition as an additional risk as it compromises other personal or professional activities of the team members.

3. *Added value to business cases*
    Prizes can act as a catalyst for business creation and growth because they make a diverse pool of resources converge and become available to entrants interested in the creation of a commercial enterprise. The added value of prizes to the commercial strategies of entrants takes forms such as the chance to win a sizable cash purse that becomes a payoff for investors or the profit of the enterprise (in the GLXP for example the Grand Prize represents a 20 per cent return on investment in a $100 million mission); publicity and reputation that help to raise funding or market technologies; valuable in-kind benefits (in the GLXP for example free access to expensive software and discount launch services); or access to professional networks and events to find new customers and partners, and gather other key resources.

4. *A wide range of opportunities*
    Prizes generally offer diverse opportunities to achieve other personal and organizational goals through different forms of participation (e.g. team member, volunteer, partner, team sponsor, friend), regardless of competitive performance. When there are low entry barriers to the prize these opportunities are available to diverse people, including students, women and individuals with no previous S&T experience and teams that are aware of their low chances of winning the prize. Prize participation represents an opportunity to, for example, generate extra income, demonstrate technological leadership, gain prestige/popularity, learn more about technology development and project

management, raise interest in science or channel ideas and energy into an exciting project. All prize entrants are to some extent motivated by these diverse opportunities.

This perspective to the incentive effect of prizes can explain the participation of a wide range of entrants and put it in terms of goal-oriented behaviors and peculiar perceptions of what prize participation offers. It is possible to identify three broadly defined groups of entrants based on their ultimate goals as either revealed in surveys or observed in their strategies. Although the goals of entrants might change over time in long-term prizes, the evidence shows that goals are generally maintained throughout the competition. Drop-outs from prizes can be explained in some cases by changing goals.

First, there are a small number of *challenge teams* whose single most important goal is to win the competition regardless of their perception of other opportunities and interests in subsequent achievements (including commercialization of technologies). These teams involve very proactive, intrinsically motivated and willing-to-compete team members, team leaders that seek prestige or fulfillment of other personal desires and/or pre-existing projects that put them closer to the achievement of the prize challenge. These entrants seek to pursue the shortest possible problem-solving path, i.e. reducing their engineering effort or shortening development lead times to accomplish the prize challenge. The idea of predicting the prize winner by looking at this group is very appealing. There have been only three (two of them unconventional) GLXP teams that more clearly match this profile (teams T4, T11 and T14) and a similar number in the AXP (e.g. Scaled Composites and Da Vinci Project) and NGLLC (e.g. Masten Space Systems and Armadillo Aerospace).

There is also a share of *startup teams* that focus on the creation of a longer-term, commercial enterprise based on the prize technologies. The key factors that explain their participation is the potential market value of the prize technologies and the value added by the prize to their business cases. Good competitive performance can contribute to their business strategies, but their R&D decisions are ultimately profit-driven even if that involves increasing engineering efforts, longer development lead times or meeting more complex design specifications not required by the prize rules. These entrants can be pre-existing companies in the prize sector, companies that re-direct their activities or teams backed by established companies. There have been at least nine GLXP startup teams, all of them conventional, including three inactive teams and two already withdrawn (for example teams T25, T8, T3 and T10) and possibly a similar share in the AXP (for example Advent Launch Services and PanAero).

The NGLLC is a particular case in which challenge and startup goals may overlap because prize victory also represented a clearer opportunity for subsequent rewards (i.e. potential technology procurement contract opportunities, something that both winners benefited from).

Finally there is a *diverse majority* of teams that find in the prize the opportunity to achieve other personal and organizational goals. These are very diverse, generally newly created teams that can easily outnumber other types of entrants if entry barriers are low because the prize represents the opportunity to work on a project that otherwise they would not have access to. They neither are as competitive as other entrants nor have probably the best chances to win the prize, but are part of the process of inducement of other outcomes of prizes that involve diverse opportunities to learn and gain experience, develop professional networks and fulfill other personal and organizational goals. Some of these entrants might eventually seek to commercialize their technologies but do not have a clear profit-oriented strategy. In the GLXP these are predominantly unconventional teams (11 unconventional and four conventional teams) (e.g. teams T13, T19, T20 and T14) and comprise most of the teams that have withdrawn or are relatively inactive in terms of technology outputs (that includes three inactive teams and four already withdrawn).

This perspective also offers explanations on the late entry phenomenon and the evolution of prize participation over time. First there is a due diligence period in which would-be entrants (and their team members, sponsors and partners) investigate the feasibility of the project and the merits of the pursuit and seek initial funding. Those involved in this process enter the prize only if they convince themselves (and others) that their participation is worthwhile according to their own criteria. Second, some entrants have longer-term goals or goals that can be achieved regardless of their prize performance and therefore their projects are not time-sensitive (or at least do not depend on the prize time line). They still enter prizes because there is an added value in participation. The lower importance of the cash purse for late entries can also be associated with this. It indicates either the acknowledgement of decreasing probabilities to win the prize (because the number of competitors increases) or the perception of other relatively more valuable benefits of prize participation. There are also competitive strategies to consider. Entrants may have a secrecy preference and wait until their technologies are more mature before announcing their official participation (Pomerantz, 2011b). In both GLXP and NGLLC there have been 'mystery teams' that decided to officially enter the competition but maintained their names and membership unrevealed for some time. It remains uncertain however how other team-level strategic factors affect the timing of the decision to enter the competition. Early entries have for example the advantage of longer public exposure and priority to access certain resources or partnerships. But in turn these early entries expose themselves to higher costs of attracting partners and investors (as the competition may still not be very visible) and the higher technological uncertainty of a problem for which the most efficient approach to find a solution has not been proposed by anyone yet. Finally there is a prize dissemination factor that we should consider. Prize sponsors may seek to sign up potential competitors during the process of prize design, but many people will still learn about the prize through regular media such as the Internet or magazines possibly months after the prize announcement. This can explain for example international late entries in the GLXP and AXP. Although these prizes have had global reach they were both launched in the USA.

We should also consider that in prizes with an extended entry period, such as the GLXP and AXP, while the incentives offered by the prize generally remain constant over time (for example the amount of the cash purse) the technological, economic or other conditions of the context are likely to change and possibly make participation more or less attractive at different points in the prize time line. This affects not only officially enrolled teams but also would-be entrants that might continue or abandon their intent to enter the competition depending on whether they perceive auspicious or unfavorable conditions respectively. In the case of the GLXP for example the unfavorable economic context may have had two counteracting effects. While it may have prevented more early entries, it may also have played a positive role from the point of view of the late entries because it prevented early entries from getting as much of a head start as they could have (Pomerantz, 2011b).

Finally from this perspective we can also explain why traditional industry players (such as the Boeing Company, for example,) and other established space companies have not entered these prizes. They have not entered because they considered that this kind of project either did not have any commercial merit or was not technically feasible (i.e. their private information indicated low or no market value or unsolvable technical difficulties). Or if they did perceive the merits and feasibility of the projects, they did not perceive any added value in prize participation. More generally prize challenges such as the GLXP mission involve a discrete, one-off product and winning the competition is not in essence a sustainable business over time.[3] Established companies have a costly structure to maintain and that requires undertaking projects that are likely to generate revenue streams through for example space hardware sales or provision of services. The prospects for a lunar commercial market (or suborbital space tourism in the case of the AXP) have been generally uncertain and unlikely to make more attractive the business plans that draw on the prize technologies. There may also be company-level strategies whereby established firms for example acquire small startups rather than developing the technology themselves. Northrop Grumman completed its acquisition of AXP's Scaled Composites after this team won the competition. Team partners may also adopt a similar rationale in their decision-making process. The GLXP project for example is technically feasible for them but their direct participation as a team is possibly not commercially viable. A few large traditional companies (for example Raytheon Company) entered the competition only through partnerships with GLXP teams to provide technologies and other in-kind support based on their existing technical solutions. Other companies, particularly the smaller ones, partner with prize teams to increase their exposure to potential new customers and develop heritage for their technologies.

7.2 PRIZE R&D ACTIVITIES

To better understand how prize R&D activities differ from traditional industry practices, this investigation probed the relationship between prize-specific conditions for R&D (i.e. fixed development lead times and the lack of up-front funding) and the designs, technology sources and extent of collaborations in prize projects (H2). In particular this hypothesis anticipated that shorter development lead times and more significant funding gaps (which result from challenges that are more expensive to achieve and the lack of up-front funding or other economic support in competitions) lead to simpler technological designs, more significant use of existing technologies and more collaborative R&D efforts. The evidence shows that prize time and funding conditions do affect the overall organization of prize R&D activities, but the kind of designs, technology sources and collaborations observed in these competitions are not exclusive features of the prize context. One of the most interesting emerging features of the aggregate of prize R&D activities is in fact the extent to which they encompass both new R&D efforts and pre-existing projects that are accelerated or re-directed. Moreover, these efforts can be widely distributed across geographical and organizational boundaries as entrants engage many other individuals and partner with many other organizations to source knowledge and other resources and outsource technology development. This R&D effort generally involves an iterative problem-solving process that entrants undertake to achieve the prize target or other related goals. Fixed deadlines and funding gaps to undertake projects become part of the problem entrants have to solve and that requires them to balance technical aspects of their solutions with the management approaches they choose for their projects. These project management approaches concern specific strategies in terms of engineering, fundraising and other procurement efforts that entrants implement during the competition. This implies that the extent to which prize designs can influence the configuration of entrant-level R&D activities is limited. Ultimately the configuration of these activities depends on the abilities and goals of the participants if we consider that prizes engage diverse entrants and their undertakings are sometimes aimed at other goals beyond the prize challenge.

Let us examine how the R&D activities induced by these three prizes relate to the conditions they set in terms of technology development lead time and funding gaps. The three prizes have concrete expiration dates but the actual development lead time available to teams varies significantly depending on the point of entry in the prize timeline. There are also different perceptions of the extent to which time and funding are actually a constraint, which ultimately depends on the goals and ability of the team. In the GLXP for example the prize deadline changed twice to allow even more time for a project that was technically feasible within the initial timeframe. Past government-led robotic missions have been more complex and accomplished in similar or shorter time frames and in the context of much more bureaucratic organizations. Still the GLXP involves a technically complex, multi-system project and creates a significant funding gap for small, informally organized teams. Reducing costs and/or gathering the necessary resources therefore becomes a financial challenge in itself, which is more essential to the GLXP than the limited lead times allowed by the prize deadline. To accelerate their projects and control costs the teams introduce simpler designs and rely more upon existing technologies, even when they do not perceive time- or funding-related constraints. Simpler designs can also help to reduce project risks that concern planetary missions under pressing time conditions. There are teams with strategies that involve the development of own technologies and in-house manufacturing. Most of the teams also lead a collaborative effort that involves partners and other collaborators. The lack of correlation between the perception of time and funding constraints and the number of partners suggests that collaborative efforts are part of individual team strategies and not a result of the prize design. There are many other interactions that are more sporadic but also important. Some GLXP teams for example have discussed technical solutions with former NGLLC teams (now space startups) or amateur rocketry associations. Prize teams also draw upon contributions of friend companies and significant volunteer effort.

Unconventional entrants in general have been more likely to introduce simpler designs to shorten development lead times and respond to a lack of funding, but at the same time these teams have been less likely to consider time or funding a constraint. This means that they have either mission time and cost estimates that are overly optimistic or a natural propensity to implement approaches to problem-solving based on simpler solutions regardless of the conditions set by the prize. On the other hand conventional entrants have a more marked perception of time and funding constraints which can be associated with approaches to space projects inherited from traditional industry practices and not only with a particular prize challenge definition. These teams are more likely to partner with other organizations to overcome those constraints, which can be explained by denser, pre-existing professional networks that unconventional entrants may not have. There are no significant differences in the extent of the use of existing technologies between types of teams.

The lack of previous technology development work on suborbital travel technologies makes it difficult to assess the effect of the AXP deadline on this kind of project.[4] It is certain that it became a challenging project for independent R&D groups and small companies, and funding emerged again as the main constraint. The NGLLC on the other hand differs in the very design of the competitions. Multi-year prizes have shorter development time frames that allow limited R&D activities, but in turn enable incremental progress and more attempts to win the prize. The NGLLC 2006 allowed only about six months between prize registration and the competition day. Those short lead times may lead to more agile forms of R&D organization and testing but also to the existence of teams that were unable to complete their projects and fly their vehicles. In both the AXP and NGLLC simplicity was also among the top design criteria. But while in the AXP this was more likely the preferred strategy to reduce costs, in the NGLLC it was the result of new, more efficient non-aerospace designs introduced by unconventional teams. In the AXP case there is evidence of diverse technology sources including in-house manufacturing, subcontracting and off-the-shelf procurement. NGLLC teams instead relied mostly on in-house, craft manufacturing. None of these prizes induced efforts as collaborative as those induced by the GLXP. There is evidence however of more significant collaboration between teams in the NGLLC than in the other two prizes. This was enabled by the race-like organization in which all teams compete and attempt to win the prize in a sponsor-organized public event. Interestingly the evidence also shows cross-prize collaborations that involved a few USA GLXP teams that discussed technical solutions with former NGLLC teams (now space startups).

Further comparisons with other instances of organization of R&D activities in the space sector show some remarkable differences with traditional space agency-led approaches but also some similitude with other

Table 7.1   Main factors that determine space R&D organization in selected contexts

	Traditional space industry	New space industry	Other private space initiatives	Prizes
Target problem	Broad agency mission's goals	Produce marketable technology solution	Accomplish personal and organi-zational goals	Prize challenge within given time and no up-front funding
Design criteria determinants	Advancement of latest technology	Market/ customer driven	Own ideas and available resources	Prize requirements, limited time and funding, entrant goals
Technology sources	Extent of in-house capability, influence of broader policies	Strategic choice, control of IP	'Accomplish as much as possible with technology at hand'	Vary significantly according to goals
Collaborations	Relationship with prime contractors, universities, other according to broader policies	Strategic alliances	Volunteers	Collaborative effort: volunteers, friend companies, partners
Example	NASA Other space agencies	XCOR Aerospace Inc. SpaceX Corp.	Copenhagen Suborbitals	GLXP, AXP, NGLLC

Source: Author's analysis based on data presented in previous chapters.

commercial sector and private space R&D instances (Table 7.1). Prize R&D activities can still adopt diverse forms of organization. The contrasting definitions of the target problem that R&D performers have to solve help explain how R&D activities are organized in each context. The most interesting feature of prizes in this regard is that, although there is a more concrete problem definition, participation is not limited to those that seek to win the competition but is also open to others that pursue other goals. Distinct goals explain very diverse design criteria, strategies for technology sourcing and collaborations. In prizes we can observe R&D activities similar to those of the new space industry if we examine entrants that seek to create a commercial enterprise (startup teams) and similar to other amateur private initiatives if we examine entrants that seek to use the prize opportunity to, for example, learn more about space technology (diverse majority). If we examine challenge teams we observe more unique features of R&D based on the minimum engineering effort criterion which actually defines an entire mission approach that generally involves doing less engineering, using more COTS technologies and buying cheaper components, and discarding 'ideal, optimal or best' solutions to focus on the prize target achievement. Startup teams are more likely to consider the market value of their technologies and choose strategic partners for longer-term projects beyond the prize target. Other teams balance their designs between prize requirements and other goals and draw more on volunteer effort in what is considered a one-off project. At the aggregate level there are remarkable differences with the traditional space industry. While space projects traditionally rely on more complex, high performance and heritage technologies, prize entrants draw on simpler and low cost designs, use existing (sometimes non-space) technologies to speed up projects, and draw much more on collaborative efforts.

We also have to consider that project costs and funding opportunities can greatly influence R&D organization in space activities. In government space missions cost has generally been a dependent variable and development lead times have been many times extended for a number of reasons. Funding in this context depends on budget approval and broader policy goals related with basic research and employment, among others. Large overhead costs of agencies and large contractors explain how expensive traditional space projects are and the lack of tension between mission schedule and costs (i.e. a faster development is likely to be less expensive as well). In other non-prize contexts cost is an independent variable and R&D performers can adjust their projects and extend deadlines to match their actual budgets and opportunities. Potentially profitable technologies help companies in the new space sector to raise funding from government and commercial contracts, private investors and venture capital. We can expect in this context a cost/schedule trade-off generally observed in new commercial product development.[5] Space amateur initiatives attain other personal and organizational goals and seek sponsorships and (monetary and in-kind) donations to support their projects. These initiatives are undertaken by low-cost flexible organizations but their plans may still fall apart if they lack some minimum support. Prizes set minimum technical requirements that entrants have to meet before a certain deadline and without any up-front economic support. Prize projects may involve high-risk technology development which makes the development of a business case based off the prize technologies more complex. Teams have to trade design features and control over their technologies to raise enough support. Fundraising is increasingly difficult for entrants with projects without commercial value or without the appropriate experience/skills to achieve the target. Technology development efforts might wither in those cases. Different goals and perceptions of the actual target problem still explain the range of effects of limited time and no up-front funding in prizes. There are small, flexible, non-profit organizations but also for-profit teams. There are also challenge teams particularly interested in increasing funding because this can 'buy time' or shorten development lead times.

The examination of team-level configurations of prize activities offers insights on potentially new-to-industry forms of organization. This research was able to identify four illustrative examples of team R&D organization (Space Agency Legacy, Universities Partnership, Partnerships Network and University Spin-off). Those examples also show that the form of R&D organization that teams ultimately adopt expresses factors such as their priority goals, knowledge, skills and available resources and may not be greatly influenced by prize designs. For example while some teams draw on existing COTS technologies to reduce engineering effort and costs, others seek to use in-house capabilities to gain hands-on experience, develop technologies for other projects or commercialize hardware. Those exemplar organizations are also evidence of generic new-to-industry approaches to space R&D. There are open source-like organizational forms (i.e. with explicit open-source strategies to coordinate distributed efforts), highly networked organizational forms (i.e. organizations that design partnership schemes to source technologies and expertise) and open-innovation organizational forms (i.e. that are open to significant knowledge flows from multiple sources). Moreover, there are several entrepreneurial- or startup-like approaches in the GLXP that differ from new space industry enterprises mainly in terms of fund-raising and revenue models. New space industry startups depend mostly on government contracts, venture capital and commercial contracts/services. GLXP teams have also raised funding from private investors and received contracts from NASA, yet their most important support comes from partners, sponsors and significant in-kind efforts (e.g. volunteers) and resources (e.g. access to specialized facilities).

More generally prize R&D activities encompass both new R&D activities and ongoing projects that converge toward the prize challenge. New R&D efforts have a wide range of starting positions in the prize timeline because of late entries. Ongoing projects are accelerated or re-directed and converge toward the prize. They have varying degrees of progress and can contribute technologies to the prize that range from conceptual ideas to prototypes or even working technologies. The caveat is that not all these R&D efforts are aimed at prize achievement, but they may still be potentially valuable to address other technological problems. The evidence also shows that entrants draw significantly on existing knowledge and technical solutions when these are readily available. There are significant adaptation efforts and improvements to use existing technologies in own projects. From this point of view this kind of user-led innovation is still a sector- and type-of-prize-specific phenomenon. In the case of the GLXP for example significant knowledge and technologies needed to accomplish the mission are readily available. Some of them are space-grade technologies and others are technologies successfully adapted from other industries. This is enabled by broader trends of miniaturization and significant reduction of costs that teams have taken advantage of. Even teams that pursue commercialization are opening up new business opportunities for service delivery mostly based on the adaptation and implementation of technical solutions that otherwise would be primarily used for government-led space programs.

Prize R&D activities can also be widely distributed across geographical and organizational boundaries. Efforts involve not only prize entrants, their partners and other collaborators but also countless volunteers, consultants and a vast array of professionals and companies that contribute knowledge and technologies to distinct parts of each project. The process even involves the public when for example teams source ideas and receive feedback through their websites, conference presentations and public talks in a kind of crowdsourcing process. All these participants can be widely dispersed around the globe. In the GLXP the teams report headquarters in 17 different countries, but the XPF estimates include team members from more than 40 countries. Moreover less than one-third of the teams organize their activities in a form that requires their members to meet regularly in the same location to work on the project. This is possible because projects have a lower level of complexity than other traditional space projects and/or entrants make very efficient use of new communication technologies and virtual collaboration tools (these technologies and tools were not available for example when the AXP was launched).

It should be noted that there are at least two factors that suggest that the collaborative efforts led by GLXP teams are competition- or at least sector-specific (i.e. the examination of other competitions may uncover different configurations). First, entrants (particularly conventional teams) use specialized partners to reduce programmatic and technology risks by using their proven technological solutions and relying upon their expertise, which is particularly important in a high-technology sector such as aerospace. Second, building a reputation of proven solutions is very important to succeed in the space business, and the GLXP gives many non-traditional aerospace organizations (particularly smaller companies) the opportunity to invest in building reputation and demonstrate technological leadership by providing technologies to prize entrants. This is a unique opportunity considering that this is a sector that has been traditionally reserved to government agencies and large corporations. Moreover, the opportunity is available not only for aerospace-related companies but also for companies that manufacture components and systems that have to deliver high performance and reliability in harsh environments other than space.

On the other hand there is a downside to this globally distributed effort. Collaborations between entrants are possibly not as significant as they were between for example NGLLC teams. This is likely to be the result of not only a more competitive environment created by the GLXP and the range of team goals but also a lack of geographical proximity among team offices and workshops that are distributed worldwide. This limits the chances of face-to-face interactions to annual GLXP summits and industry conferences. Not all teams however participate in the prize annual summits and even fewer coincide at conferences. Similarly for at least three years the GLXP official website offered no tool for more intense virtual interactions and collaborations between teams. There is still some evidence of a handful of teams that connect sporadically and informally to discuss technical aspects of their projects. The NGLLC which held annual races with participation of all teams created many more opportunities for interaction and collaboration between teams.

Among the most interesting features of the iterative R&D process involved in prizes is also the convergence toward the prize target. There is a hierarchy of decisions, from overall approach decisions to smaller technical choices and feedbacks that result from previous decisions, information exchanges and interactions and dissemination of project information. Entrants also perform significant trial-and-error and prototyping iterations to find solutions to specific technical problems when they do not have access to the most appropriate parts or components, seek to introduce new ideas and experiment before making subsequent decisions. An interesting feedback loop in the case of GLXP projects is the technical impact of fundraising efforts. In the course of these R&D activities the teams perform significant fundraising and business development efforts to finance their missions, which sometimes impose new design requirements and consequent efforts. The successful commercialization of payload transportation services for example generates revenues for the team but in turn increases the spacecraft mass and eventually the launch costs. Other revenue models such as those based on data sales do not impact spacecraft designs, but require accomplishing the mission first.[6]

This process also encompasses key trade-off decisions. Hard choices or decisions have to be made to solve 'contradictions' in technical aspects (e.g. costly own development of the desired technology v. adaptation effort of cheaper COTS component) or overall approaches (e.g. less expensive project v. increasing fundraising efforts). In the GLXP this is more noticeably in the decisions about buying a commercial launcher or developing a launcher. Most of the teams ultimately seek to use proven, commercially available solutions because of the shorter prize time frame for additional engineering effort. Commercial solutions however impose both minimum funding requirements and maximum payload capabilities and the decisions for their selection ultimately influence the design and development of spacecraft and the rest of the systems required to accomplish the mission. Entrants may attempt to reduce the mass of spacecraft and revise their decisions and use cheaper launch vehicles, but this might be actually counterproductive in space projects because smaller components are usually more expensive as well.

This problem-solving process is certainly not linear and is influenced by changing perceptions and new opportunities open to teams. In the course of their activities entrants come across new opportunities, for example to commercialize their technologies and services or pursue related projects. Choices in this regard affect their R&D activities and the ultimate outcomes of the prize. Changing perceptions about the prospects of winning the competition or finding support to continue the projects can also change goals and activities of entrants.

7.3 PRIZE TECHNOLOGIES

This investigation anticipated different motivations to enter the prize and linked the characteristics of the prize technologies to the characteristics of the prize entrants and the technology incentives related with the prize challenge (H3), novelty and maturity being the dimensions adopted to describe those technologies. The evidence shows that the technologies developed by unconventional entrants might be more novel than those developed by other entrants but are also more experimental or aimed at technology demonstration (that is, controlling for the time of involvement with the prize and pre-existing projects). In other words most of those technologies are not yet ready for mission use and require further testing and development. Considering that novelty is not among the top design criteria, these technologies may be the result of increasing experimentation and trial-and-error iterations as these unconventional teams draw upon larger volunteer efforts and are less risk-averse. Other less expected sources of new concepts are science fiction and non-space industry designs. Among these groups of unconventional entrants, challenge teams appear much more active in technology development and potentially more innovative. Conventional entrants on the other hand also introduce novel technologies but these are generally closer to actual mission use or commercialization. The evidence indicates that a few novel concepts introduced by these teams relate with the advancement of technologies available from partners for commercialization of lunar services. These entrants not only have access to more knowledge and partners but also prioritized the creation of a longer-term commercial enterprise.

Whether technological advancement reaches implementation or commercialization phases ultimately depends more on the technological target implicit in the prize challenge definition and the market value of the technologies. Among these three competitions the GLXP appears as the prize that targets technologies that are closer to commercialization in the innovation pathway not only because of its challenge definition but also because of the potential markets associated with it. The competition per se does not require commercialization but its challenge addresses a technological problem that is associated with a potentially sizable market of lunar technologies currently driven primarily by space agencies and large companies.

Diverse entrant goals influence the technologies developed in prizes as well. The evidence on prize technologies shows that the effect on the advancement of technologies is manifest in not only the winning entry but also the wide range of other intermediate technologies advanced by entrants. These outputs are contributed from the time of the prize announcement, by both entrants and their vast networks of partners and collaborators. The opportunity to investigate an ongoing competition such as the GLXP has helped to examine those technologies in more detail and identify the kinds of outputs described earlier (in a retrospective analysis of competitions such as the AXP and NGLLC it is more difficult to make that kind of examination). Overall the aggregate of technological activity in the GLXP has been mostly at medium-high levels of maturity (somewhere between TRL 6 and TRL 8) and the contribution of entrants has been diverse. Most of the unconventional teams are part of a diverse group with goals beyond the prize target and technologies that span across both scales, innovativeness (ranging from current-day to novel technologies) and implementation (ranging from research/experimental purposes to actual implementation or commercialization). Only half of this group of teams has contributed significant outputs in the form of technologies or new project management approaches to space missions.[7] A few unconventional entrants that have been identified as challenge teams can be associated with more creative solutions to specific problems, more entrepreneurial orientation and also development of technologies at medium and high TRL. These outputs can generally be associated with efforts to shorten typical development lead times and reduce project budgets by introducing simpler, more affordable and efficient solutions (such as new solar panel-antenna configuration, internal mobility mechanisms and new approaches to rover mobility). A few potentially commercially valuable new technologies have also been introduced by these teams (e.g. new software algorithms to control rovers' systems). The work of conventional entrants on the other hand is associated with the advancement of technologies at higher maturity levels, for mission use or commercialization of services such as payload launch and delivery. In the GLXP a group of startup teams mostly formed by conventional teams can explain why these entrants focus on the development of technologies at higher maturity levels for commercialization.

A closer examination helps us to better understand the range of prize technologies along the innovation pathway and identify four main types of technology outputs produced in three years of competition:

1. *New concepts and experimental technologies*
   These outputs either contribute new-to-industry concepts or solutions (e.g. new mobility concepts such as ball-shaped robots, new software algorithms) or advance the maturity level of existing technical concepts (e.g. hoppers or leg-enabled systems). Some of these may not reach production/implementation phases in the foreseeable future yet others may become truly breakthroughs in planetary exploration or other industrial applications. There is not strong evidence to establish a causal connection between these outputs and the design of the prize competition but an open-ended definition of the prize challenge certainly enables their application to the GLXP mission.

2. *Creative solutions to specific, well-defined problems*
   These outputs address specific technical problems such as operation in the harsh lunar environment and balance of mass/size/capabilities of spacecraft. These are problems already faced in previous planetary missions and are well-known to space engineers. These solutions are still unproven, generally introduced by unconventional teams and likely to be first implemented in GLXP missions (e.g. solar panel-antenna integration, internal mobility mechanisms). These solutions are the smart response of teams to solve those problems cheaply and quickly.

3. *More affordable and simpler versions of existing systems*
    These outputs comprise mostly landers and rovers based on existing, more mature technologies (e.g. wheeled rovers) and are aimed at the achievement of the prize challenge or delivery of payload services. The ultimate purpose of these systems determines their quality: faster prize challenge achievements require engineering efforts that satisfy only the minimum requirements of the prize (e.g. a rover designed with a life-time to traverse 500 meters only); commercially oriented space-craft development requires other kinds of engineering efforts (e.g. a rover designed with payload capability). These technologies share similar attributes in teams with similar goals and can be considered as *me-too systems* rather than original developments.
4. *New development and mission approaches*
    These outputs comprise actual project management methods and mission approaches that involve commercialization of payload delivery service, consulting services (by the team members individually or as a new enterprise of the team) and the provision of mission support services. These are generally new approaches that combine partnerships and other forms of collaborative effort, more affordable technologies and new business models. In part these outputs are associated with the schedule/funding requirement conditions posed by the GLXP challenge, as explained later.

The evidence suggests that technology development activities in the AXP were focused more on medium TRL. The winning entry (the most notable technology output of this competition) was a spacecraft that, although based on existing technologies and materials, involved the development of a novel configuration and novel uses of those technologies and materials. Other entries worked on more experimental approaches to the achievement of suborbital flight such as balloons, aircraft with novel configurations and vertical take-off vehicles. Again in this case most of the more novel designs were in early phases of the innovation pathway and were not more than conceptual ideas rather than actual technologies and functional prototypes. In the NGLLC, with a more specific target, technology development focused on medium or medium-to-high TRL and converged toward similar approaches but with different levels of performance. The technologies in this case comprise for example new controls, rocket engine components and operational capabilities.

The evidence on the range of technologies entrants work on is not only confirmation of the ability of prizes to induce the development of alternative solutions to the challenge but also their lack of means to focus problem-solving efforts on specific targets without hindering innovation.

An example of the diverse solutions contributed by entrants is the GLXP with development lead times of missions that range from 30 to 90 months and expected mission costs that range from $5 million to $100 million. This range of mission approaches includes for example approaches that comprise both the development of own launcher and use of commercially available launch solutions. While this is a potential source of new project management and mission approaches it is also the result of much R&D activity in the pursuit of goals beyond the prize target. This is in part because, unlike more focused prizes such as the NGLLC, the GLXP neither requires creating a spacecraft of certain kind and demonstrating its capabilities nor specifies how the prize challenge has to be achieved. More strict requirements in the GLXP (such as a requirement for certain spacecraft mass in landers and rovers) would have possibly led to less variation in the range of mission approaches.

More generally, the examination of these three case studies suggests that the characteristics of the prize technologies are a function of not only the prize challenge definition but also ongoing projects, current-day technologies and entrant-level factors. In terms of prize challenge definition there are four key design parameters that influence outputs:

1. *Technical definition of the prize target*
   This represents the technological gap that entrants have to close to be able to claim the reward and is a concrete measure of achievement or standard by which the performance of entrants is measured. In the case of the GLXP for example this target was defined in terms of landing, mobility, mooncast transmission, data uplink and payload. This is what drives most of the activity of entrants that seek to win the prize before anything else.

2. *Type of prize according to required output*
   There are prizes that explicitly require building new systems and demonstrating their capabilities (i.e. technology demonstration prizes) such as the AXP and NGLLC cases, and others that let entrants decide about the means to accomplish a certain technological feat (i.e. prize for technology-based achievement) such as the GLXP case. In this manner sponsors can set technical requirements that focus efforts and increase the likelihood of obtaining a winning entry that is compatible with their expectations. That may however limit the introduction of more creative, affordable or efficient methods and technologies that the sponsors are not aware of. Interestingly the latter may be still preferable in some circumstances. If sponsors do not use the requirement of developing a new technology with certain characteristics, entrants may win the competition 'by cheating' if they find an

unexpected solution that meets the (ill-defined) prize requirements but does not add any value to current-day methods and technologies, that is, without producing any technological development or innovation.

3. *Prize expiration date*

   While prize targets that involve significant technology gaps and challenges of expensive achievement require longer lead times (to undertake R&D or raise funding respectively), shorter development times may induce the implementation of more efficient problem-solving approaches. Technology development still requires in any case a minimum development lead time that prizes have to allow to come up with a solution to the challenge, time that might be significantly longer when prizes involve pure research at lower maturity levels. In these prize cases the development of new technologies (i.e. AXP) involved a few years of work, and increasing efficiency and incremental development (i.e. in each NGLLC competition) have been made within months. In the GLXP, technology and services commercialization started as soon as the prize was launched and teams started to seek funding.

4. *Cost of achievement*

   Prize challenges that are difficult to achieve because they involve significant R&D efforts may induce for example the advancement of technologies with lower maturity levels if that implies finding more affordable solutions to the prize challenge or less expensive project management approaches. But if too much funding is required to achieve the prize challenge (because for example the prize requires advancing technologies from pure research to commercialization) and there are not favorable conditions to raise funding, prizes may not induce efficient problem-solving efforts. In that case entrants would have to divert most of their efforts to raising funding, and actual R&D activities would be hindered. Entrants may even withdraw from competitions or remain inactive when funding is a constraint for them. The alignment of the prize target with potentially sizable markets for the prize technologies and/or a prize challenge that includes a commitment to buy prize technologies can offer an additional incentive to advancing technologies toward implementation/commercialization stages at higher TRL or help in fundraising.

The evidence also contributes insights on the connection between prize technologies and ongoing industry projects and readily available technologies. Prizes can provide an additional incentive to accelerate ongoing projects and continue/restart industry projects that were held back for diverse reasons. There are examples in these three prizes. A fair share of GLXP teams report that their prize projects are the continuation of previous projects. Some evidence indicates that Scaled Composites might have started its space-related projects before the company entered the AXP. Masten Space Systems entered the NGLLC with technologies the company already had under development. The former AXP entrant Armadillo Aerospace entered the NGLLC to continue with its space projects. Prize participation also offers an opportunity to try new designs that were conceived before prize announcement. This is the case of at least one team in the GLXP. All these examples suggest that prize projects are not always new R&D instances that end with the achievement of the prize target but sometimes the continuation of pre-existing projects or the beginning of new ones. Prior knowledge and current-day technologies that are relevant to the prize challenge also enable subsequent developments and directly contribute to the projects of entrants. In the GLXP about 45 per cent of the teams use technologies that are somewhat new and another 30 per cent technologies that are not new at all. Knowledge codified in manuals, team member expertise, expert advice and parts and components (including non-space technologies) are for example inputs in GLXP projects. In particular commercially available COTS technologies help GLXP entrants to speed up, reduce the cost of and attenuate the risks of their projects. But most importantly existing knowledge, expertise and technologies enable the pursuit of a prize target that is closer to actual implementation or commercialization in the innovation pathway. In the case of the AXP, although the idea of space tourism was not new when the prize was announced, knowledge and relevant technologies were limited to studies and scholarly literature on the topic and (at least in the case of the winning entry, which draws on this previous experience) experimental research undertaken by the USAF/NACA X-15 program. The NGLLC involved building and flying a VTOL vehicle with relevant technologies readily available, including knowledge and expertise from NASA Apollo landers and the 1990s antecedent Delta Clipper program.

Finally entrant-level factors such as goals, skills and resources available to entrants also influence the technologies they work on. The evidence from the GLXP case is illustrative in this regard. To win the competition entrants seek to perform the least possible engineering effort by either (a) working on advancing technologies that are already closer to flight-ready or (b) adapting/using existing, proven technical solutions. Since the GLXP does not explicitly require building any specific system, the choices about those two options depend on whether technologies and/or funding are available to teams. When facing a lack of funding teams can partner with other organizations to source/develop technologies or explore a more creative, less expensive way to solve technical problems. While the latter is associated with the introduction of more novel ideas, experimentation and lower TRLs, both (a) and (b) are linked to higher TRLs. Other teams that have goals beyond the prize (e.g. creating a company, having fun, other organizational goals) do not have to optimize their prize performance in that way. These entrants may seek to explore novel technologies, use existing solutions or develop technologies with commercial merit, which involves R&D in a wide range of maturity levels. Decisions in those other cases are ultimately determined by specific goals and require individual entrant-level analysis for their understanding. In the GLXP case this diversity comprises experimental tests of alternative propulsion methods, novel mobility systems concepts and landers with characteristics not required by the competition (e.g. scalability or modularity) among others.

Two final considerations apply to this analysis. The GLXP is an ongoing prize competition and therefore the number of significant prize technologies is likely to increase over time. This consideration is more relevant to the possible implementation of novel technologies. A hint of the near-term developments is given by the fact that, after three years of competition, similar proportions of both unconventional and conventional entrants have team members exclusively dedicated to business development or commercialization of the prize technologies. This suggests that more technologies might eventually be advanced toward commercialization. It should be also considered that the assessment of the level of maturity of the technologies has been made at the level of the technology subsystem/system and is not detailed enough to capture the full range of possible new technologies at the level of, for example, materials or components. That might be a reason for the small number of novel technologies found in this case. Considering the competitive context, the lack of up-front funding and the minimum possible engineering effort criterion, it seems less likely (but not impossible) that teams will perform research to discover new basic principles or develop new materials.

7.4   THE OVERALL EFFECT ON INNOVATION

The most important question addressed by this investigation is whether prizes spur innovation over and above what would have occurred anyway. This investigation anticipated that prizes do induce innovation and that their ability to do so increases with more significant prize incentives, more significant technology gaps and more open-ended challenge definitions (H4). Significant evidence supports this hypothesis but there are some caveats. Increasing non-monetary prize incentives has the greatest incentive effect on entrants, including late entries, which leads to increasing

R&D activities and more technology outputs. The creation of these incentives however depends on a positive evolution of the competition and the characteristics of the industry sector and technology markets. Entrants that seek to enter the field through technology commercialization for example are attracted by the promotion/reputation value that competitions acquire over time, which is also linked to sizable, but still uncertain markets for the prize technologies. Opportunities to learn, gain experience or pursue other goals also attract many more and more diverse entrants, but these may not necessarily contribute significant technology advances. The effect of monetary incentives seems to be more indirect yet still important. They can help to disseminate the idea of the prize to broader audiences that include the media and the public, which ultimately may attract more prize entrants. The weight of the cash purse in the decisions to enter these prizes has still been relatively low compared to non-monetary incentives and technology incentives. More significant technology gaps on the other hand induce entrants to seek more creative and low cost solutions in a context of limited time and lack of up-front funding. The evidence suggests however that technological breakthroughs cannot be directly induced but only enabled by prizes. This is because entrants that seek to win the prize seek only achievement and not necessarily novelty, efficiency or high quality. These teams seek to perform the smallest possible engineering effort and that involves using existing technologies and developing technologies that are simpler and cheaper. Although these may be valuable attributes in some circumstances, technological breakthroughs may require more significant, long-term R&D efforts and additional incentives or economic support.[8] Moreover, in the GLXP, a significant number of the innovations self-reported by entrants were developed by chance, and novelty is not within the preferred design criteria. This suggests that much of the innovation activity results from processes of using and adapting existing technology and not from an actual search for innovative solutions to the challenge.

Open-ended challenge definitions also enable innovation in prizes because they offer incentives for and reward the development of a wider range of technologies. Prizes with broadly defined targets capture the attention and allow contributions of a wide array of R&D performers with diverse goals. This more diverse activity is more likely to induce an unexpected, truly innovative solution. Conversely, strict challenge definitions to deliver technologies with certain specifications might have induced a more focused effort at the cost of less diversity in approaches and proposed solutions. That kind of focused definition might pass over potentially innovative solutions that do not necessarily meet the prize requirements and potentially reduce the interest in prizes. The GLXP's challenge definition in particular is very open-ended and has enabled the innovation processes described earlier. Moreover, entrants have reported innovations in the form of new products and components that are useful for both the prize and other projects. This type of open-ended challenge definition is a necessary condition to allow alternative solutions to technical problems and alternative approaches to find them, yet it is not an innovation driver in itself. In other words open-ended challenges do not determine for example whether technological outputs are successfully introduced or commercialized. Eventually there is a limit in which further relaxation of the prize requirements may result in unfocused R&D efforts and have a negative effect from the point of view of prize sponsors interested in procuring certain technologies.

There are other parameters in prize challenge definitions that can affect innovation. On the one hand longer lead times allow more research and proof-of-concept work to advance lower maturity technologies. On the other hand shorter lead times can induce new, more efficient approaches to R&D that involve for example rapid prototyping and faster turnarounds. The most appropriate balance seems to be in lead times that are as short as possible but allow a minimum lead time required for both technical work to address the prize target and team formation and growth (in the GLXP, projects that started on average four months before teams' official entry date suggest formation periods of a few months). Prizes also have to pose challenges that can be met with resources potentially available to entrants. The R&D process may be halted if the achievement of the prize challenge is very expensive for small, independent research groups, unless additional support is provided to entrants in the form of, for example, research grants or in-kind support. To some extent only, expensive challenges may induce new forms of project organization, as seen in the GLXP low overhead cost organizations.

Overall the evidence suggests that these prizes induced innovation over and above what would have occurred if the competitions had not been held. The GLXP has induced increasing R&D activities and the development of technologies that might become significant innovations. Most of the teams are newly formed and several of them produced significant technology outputs. Moreover, the GLXP has been the main reason for half of the teams to engage in this type of project. A number of them have implemented those technologies in their missions or seek hardware/services commercialization. The GLXP also represents an opportunity for many individuals and teams to work with space technologies and potentially enter the space sector. The prize accelerated and expanded existing projects or even helped to re-initiate projects similar to those started more than 10 years ago (e.g. LunaCorp and Astrobotic). Moreover, more than

80 companies, universities and NGOs have directly or indirectly participated in (and learned from) technology development through partnerships
with GLXP teams. The teams themselves engaged at least 500 people
during 2010. Most importantly, the GLXP led to the creation of networks
of partnerships and collaborations that will transcend this competition.
The global GLXP R&D expenditures are still uncertain to this investigation and difficult to calculate with accuracy anyway, but teams might ultimately invest up to $465 million in their projects during this competition.[9]

The AXP and NGLLC also induced increasing R&D but, based on the
author's assessment, much of the activity observed in those competitions
(particularly in the AXP) was related to ongoing projects rather than new
R&D. Most of the entries to these prizes were however newly created
independent or startup teams. Third-party estimates suggest that the AXP
may have induced investments of up to $400 million (Schroeder, 2004) and
the NGLLC investments of up to $20 million in four years of competition
(Courtland, 2009).

A word of caution is required to understand the effect of these prizes.
The evidence suggests that the increasing incentive effect of prizes, more
widespread participation and increasing R&D activities do not all translate into technological innovation that addresses the prize target, because
a significant part of these R&D efforts may be aimed at other individual
and organizational goals. Very diverse entrants with distinct perceptions,
strategies, skills and resources advance technologies that are related to
the prize challenge but also have other undertakings that help them to
pursue other goals. Ultimately R&D effort levels and prize technologies
depend on those entrant-level and other contextual factors. This range
of causal factors involved in the prize process suggests that the incentive
effect of prizes is likely to vary across technological fields and is difficult
to anticipate by prize organizers.

The opportunity to investigate an ongoing competition also helped to
uncover more dynamic aspects of the prize process. Changing perceptions
and a context that evolves can re-direct the activities induced by long-
term prizes. There are also feedback effects that are not likely to affect the
causality patterns described in this chapter but can intensify or moderate
the ultimate effect of prizes over time. Increasing R&D activities (in magnitude or visibility) induced by the prize, for example, may attract more
entrants or partners to the competition. Certain technologies developed
by teams may change the perceptions of would-be entrants about the feasibility of the project. Other events such as industry's technological breakthroughs, economic crises or (more specifically) space government mission
achievements can alter the prize process significantly.[10]

A more in-depth examination of the innovation process of the GLXP unfolded other interesting features. This prize has induced overall R&D efforts at medium-to-high maturity levels and pushed developments further toward commercialization, with less emphasis on early points in the innovation pathway (e.g. basic research). Some factors anticipate what kind of innovation is likely to occur in the GLXP. The fact that teams tend to use existing launch solutions suggests that potential innovations are more likely to occur in new or advanced payloads (including landers and rovers) and not in launch technologies. Despite the use of large proportions of COTS technologies and the criterion of the smallest possible engineering effort there may still be innovation, for example in testing and adaptation efforts to use those technologies. New development and mission approaches are likely to be the most innovative contribution of the GLXP to this field (which is incidentally one of the targets of the sponsor for this prize). At the entrant- or workgroup-level there is also a myriad of R&D efforts or research threads that address diverse technical and project management issues at different points of the creation-to-implementation innovation pathway. Countless intermediate outputs and activities undertaken in competitions are evidence of that vast array of R&D activities which are generally fairly unevenly distributed across entrants.

This closer examination of prize activities also shows that the range of prize technologies is so wide and the activities induced by the prize are so diverse in essence and purpose that it is difficult to speak of duplication of effort. Moreover, most of the teams maintain that they come up with own design ideas and do not imitate or get inspiration from what other teams do. One could argue that there is duplication of effort because several teams are working on the development of more affordable wheeled rovers. But it is difficult to find similarities beyond the more visible four-wheel or six-wheel traditional configurations. Factors that lead to the same conclusion include the number of prize entrants and other participants with diverse backgrounds and interests. There are also multiple back-end business markets that can be supported by the technologies developed in pursuit of the prize, an open-ended challenge definition and relaxed criteria of eligibility to enter the competition.

These case studies show that innovation in prizes is not an isolated process but relates to industry developments and the structure of the industry sector. Entrants build not only on pre-existing capabilities, knowledge and technologies but also on industry experience and ongoing industry projects (e.g. NASA's Common Bus program). There are even pre-existing R&D teams that enter prizes. Particularly in the GLXP, the teams also benefit from the experience of a number of planetary missions and from past, similar mission attempts in the private sector (e.g. LunaCorp, BlastOff!). Prize entrants also benefit from expanding partnership networks. The technologies developed by partners offer prize entrants new opportunities and possibly an advantage to win the prize or accomplish other goals. There may also be companies and other independent groups that work on similar projects but do not enter prizes. Those unofficial competitors can bring more dynamism to the technological field of the prize and make the prize projects more credible and attractive as well. In new technological fields or emerging industry sectors however there may be no R&D performers other the prize entrants. To the author's knowledge no GLXP-like mission is being undertaken by private entities other than the GLXP teams (although there might be for example a withdrawn GLXP team that has secretly continued its project). The AXP was launched when the new space sector was just emerging and only a few small companies addressed a similar target. The NGLLC targeted a niche traditionally occupied by large companies but also in the context of an emerging space sector with a handful of companies interested in the same technologies. In longer-term prizes changing perceptions are likely to influence prize developments as other non-prize R&D performers also push technology development and the industry sectors evolve. Broader trends and practices in the new space industry and other private space initiatives can offer new solutions to the prize challenge or related technical issues, and influence decisions of prize entrants and investors. GLXP teams in particular benefit from more general changes in beliefs regarding the potential for private space initiatives which might attract more investors and private customers. Hence the effect of these prizes depends to a great extent on the special characteristics and idiosyncratic aspects of the space sector.

The sizable potential market value of space technologies also makes relative the importance of the cash purses in challenges that involve some kind of implementation or commercialization. How the markets for the prize technologies actually work ultimately influences the extent to which that value is actually realizable. The demand-side in the space sector comprises few actors. Commercialization in the GLXP for example is mostly driven by NASA's acquisition of data and funding for hardware demonstrations. In this case increasing demand would not only have attracted more entrants (through the perception of the benefit of reputation offered by the GLXP) but also induced a more intensive R&D effort and collaborations from partners. Private investors would be more willing to invest if there were stronger commitments from NASA or other space agencies to fund space initiatives.

Immediate commercial opportunities have been limited during the development of these competitions. Moreover, this commercialization of prize technologies in the short-term does not necessarily involve their broader adoption and dissemination, particularly if only a handful of teams are able to market their technologies (this may still be a signal of potentially good prospects for other well-performing teams). The AXP was also linked to the promising suborbital manned space flight sector, but in this case the market value of the prize technologies may have been even more uncertain than in the GLXP. Scaled Composites was able to commercialize its technologies immediately after it won the prize through a contract to deliver a fleet of spacecraft. But as of today, more than six years after the competition ended, the suborbital tourism segment (possibly the most valuable in this market) still does not serve customers regularly. The technologies involved with the NGLLC may not have had market values as significant as in the other prizes, but the opportunities for contracts and technology commercialization appeared more concrete (and they were realized in some way with contracts for further technology development).

These prizes might also influence (positively or negatively) subsequent technological developments and industry decisions. The extent of the connection of these prizes with and influence on the space sector varies. It may occur that a few entrants have a key role and dominate the dynamics and ultimate achievements of the competition, and acquire enough visibility to potentially set industry standards and influence investment decisions. In the GLXP the visibility offered by the competition to the achievements of the teams may also have significant effects on further innovations and industry decisions. NASA has already invested to procure mission data from a handful of teams. In the near future the achievement of the GLXP challenge with, for example, a hopper or leg-enabled system rather than a more traditional wheeled-rover, might influence assessments on what the most efficient technologies for planetary exploration are. Prize solutions might not be the most affordable or valuable from the commercial standpoint, but can still potentially be the most visible and most able to signal, correctly or incorrectly, what methods, technologies and markets industry players and investors should aim for. The AXP mostly highlighted the achievements of Scaled Composites. More advanced, new versions of the cargo aircraft-launcher/spacecraft configuration created by that company will be soon introduced by the company Virgin Galactic to offer the first regular suborbital tourism flights. Simultaneously other companies that did not enter the prize have developed similar technologies and followed an alternative path that involved contracts or traditional new product development. XCOR, for example, a private rocket engine and spaceflight development company, has developed a reusable, single horizontal take-off and horizontal landing vehicle, also with human transport capabilities (though in this case, for only one passenger). These two alternative configurations may have more or less equivalent capabilities and efficiency, but Scaled Composites' design benefits from the exposure given by the AXP. This is the kind of visibility that might influence investment decisions in the sector. The evidence suggests that the very decision of Virgin Galactic to acquire spacecraft from Scaled Composites is the result of that and one of the outcomes of the prize. The NGLLC was dominated by Masten Space Systems and Armadillo Aerospace, which was not a surprise for experts who considered those the teams with the greatest potential to win the prize. Later on both companies became referents in this emerging sector. Other teams also developed and tested vehicles in those four years of competition, but without the attention that the winners attracted from the media and the public. There are other companies that serve (or can potentially serve) markets that use VTOL vehicles but in this case too the visibility given by the prize may have influenced technology development decisions. NASA for example offered contracts to both prize winners. There is also evidence of GLXP teams that have consulted with or analyzed the designs of those NGLLC teams to inform decisions in their projects.

Although prizes may be able to influence industry and investment decisions by showcasing the winning entry (or other significant prize outputs), they are not necessarily able to signal the value of the technologies. This research has not gathered positive evidence in this regard and the analysis of the context suggests that the perceptions of the industry sector in relation to the value of the prize technologies are not necessarily affected by the announcement of the prize.

Finally the broader economic context can also moderate or intensify the process triggered by prizes. Favorable economic conditions help entrants to raise funding, and good prospects encourage investors to support prize teams. Conversely unfavorable conditions may hinder the prize process, particularly if it involves technologies with uncertain commercial merit, which is to some extent the case of the GLXP. The economic slowdown that started in 2007 has affected the activities of teams and delayed their projects. Certainly, the high return rates expected by private equity and venture capital to fund space projects have not been met by the economic situation. Similar phenomena occurred in the case of the AXP with the economic slowdown that started in 2001 and the risks related with NASA's Space Shuttle tragedy of 2003. On the upside, however, the GLXP has kept the R&D focus of teams that managed to attract sponsorships and other in-kind contributions from multiple sources and sectors to fund their projects. This suggests that prizes might be useful to maintain certain levels of R&D activity and innovation in contexts of recession or business contraction cycles. Considering that some teams have entered the

GLXP even in a context of economic recession, prizes might also help in recovering stagnated niches. It should be borne in mind that even with a favorable business context, the interests of potential customers and investors might be in different market segments and not exactly in the prize technologies (markets might be interested in low orbit technologies and not in planetary exploration for example).

Prizes can also induce other later outcomes beyond immediate technological outputs. Those outcomes are however more difficult to anticipate than R&D activities and technologies involved in competitions (which are uncertain at the moment of prize announcement). The evidence shows, for example, that prizes offer individuals and organizations the opportunity to learn and gain experience, which eventually might lead to technological progress. Prizes might also inspire and demonstrate technological possibilities if entrants come up with disruptive ideas or achieve challenges that were considered unachievable or very difficult at the time of the prize announcement. This potential ability of prizes is related with shared beliefs regarding those possibilities and the spirit of competitions. The organizers of the AXP for example were inspired by the early aviation prizes and purposely designed the prize to change public perceptions about the capabilities of the private space industry. The connection of this challenge with the possibilities of opening the space frontier to human activities may have helped to attract broader audiences interested not only in space tourism but in space travel in general. Extensive global media coverage may have been key to disseminating progress updates and the general spirit of the competition. The competition has been considered successful because of such extensive coverage and subsequent investments in the private space flight sector (Maryniak, 2010). Other prizes might not be able to produce this kind of effect when they involve technological achievements that are not apparent to the general public and competitions that seem not very exciting. This is to some extent the case of the NGLLC which involved a target based on a less obvious, more technical flight precision achievement and had few entrants each year.

From this perspective the prize process can be understood as a process of idea development/dissemination and transformation of the perceptions about technological possibilities, which might trigger further innovation processes and inspire other human endeavors. The case studies presented by this investigation help to reveal this process in modern prizes. The starting point is the idea to launch a prize. The prize organizer shares the idea with experts to assess its feasibility and potential and designs the prize. The organizer's beliefs about the merit of the pursuit are then stated implicitly in the spirit of the competition or explicitly in the prize rules and actions to promote the competition. The organizer also contacts potential entrants to discuss their interest in an eventual competition, 'trades' prize design features to make it even more interesting to potential entrants and announces the prize to the media. At this point only a small number of people know about and perceive the merits of the upcoming competition. After the prize announcement new entrants convince themselves of the merit of this project, but most importantly the media begin to support the idea and help to disseminate the vision and spirit of the competition. This is a very important time period in which teams have to convince new members, sponsors, partners and investors (and even their friends and family) about the merit of their projects. Promotional actions help considerably during this process to further disseminate the progress of participants and engage broader audiences. These competitions have generally been followed by thousands of enthusiasts who are interested in either specific, favorite teams or space development in general. Each new technological accomplishment that is then showcased helps to change the perceptions of what is possible in this industry. By the time the prize is won the very accomplishment of the prize challenge (if apparent and notable) becomes a clear demonstration of the technological possibilities in the field, with a potentially widespread effect on industry, the media and the general public.[11]

NOTES

1. It is worth noting that incentive mechanisms other than prizes also build reputation in this industry, although not based on visibility or publicity. For example, hardware demonstration contracts awarded by space agencies help awardees to increase credibility and raise more funding (Culver et al., 2007).
2. There are several other examples of this. In the GLXP there are teams that are not afraid of losing the time and resources they invest or embarking on ineffective technical approaches, even though the XPF sought to regulate entries and accept only those projects that were considered credible or reflected the entrants' knowledge of what they were getting in to. In the AXP and NGLLC there were also important regulatory risks related to the hurdle of obtaining experimental permits for space flights.
3. Some teams might seek to participate in multiple prizes to generate a revenue stream over time. To the author's knowledge, no participating team in these three prize case studies has participated in more than two competitions in the 15-year period covered by this investigation.
4. For reference only, the X-15 rocket-powered aircraft project that the winner of the AXP used as an inspiration for its design required about five years between contracting and initial flight.
5. This kind of trade-off has also been observed in NASA's recent smaller missions by the specialized space literature (Dornheim, 2003) and has been discussed by the new commercial product development literature (see, for example, Mansfield, 1988; Gupta and Wilemon, 1990; Swink et al., 2006).
6. Some kind of interrelation between budget and R&D activities exists in other contexts as well. For example, changes in project goals and actual developments are introduced

in agencies' space programs due to both the uncertainty about the funding available for a program and the constraints resulting from annual revisions of budgets (Bitten, 2008). Other amateur space initiatives also depend greatly on fundraising, but the interrelation in those cases might not be as strong as in prizes since they are not in a competitive, race-like context nor do they seek to commercialize technologies. On the other hand the much discussed general interrelation between corporate R&D management, investment, and debt policies depends on very different factors that include, for example, predictability and movement of debts, profits, and cash flows (see, for example, Hall, 1992; Himmelberg and Petersen, 1994; Bond et al., 1999; Hall, 2002).

7. That includes technologies such as alternative propulsion systems, new controllers using non-aerospace components and new video/navigation systems, and mission approaches such as open-source. These teams have also worked on more affordable and simpler versions of traditional designs for systems such as landers and rovers.

8. This is in line with accounts of other early and modern prize cases that needed longer lead times and/or additional support to induce significant technological advances. For example, John Harrison, the winner of the historic Longitude Prize, received different amounts of money over 25 years until he completed the most efficient version of his chronometer to measure longitude at sea (Sobel, 1996). The more recent DARPA Challenges provided seed funding to 11 qualified entrants in the form of a competitive proposal with awards up to $1 million each, dependent on performance. This helped some of the best performers, particularly those with smaller teams, to remain in the competition (Whitaker, 2010).

9. This is a very optimistic figure. It is calculated based on the mission budgets reported by seven teams and an average of $15 million for each of the remaining nine teams with significant outputs that entered the GLXP before December 2010. The $15 million average is the minimum mission expenditure expected by the XPF when the prize was announced (XPF, 2008a).

10. According to the GLXP rules, for example, the prize expires if a government funded project accomplishes a similar mission.

11. Patterns like this were also observed in the most popular, historic prize competitions. The Orteig Prize for example did not attract participants for about five years after its first announcement in 1919 and it was then renewed for a further five years. A number of attempts to accomplish the prize challenge started to get the attention of the media and when it was won in 1927, the prize sparked a widespread interest in aviation and follow up investments in this industry (O'Sullivan, 2009; Kessner, 2010).

8. Theory, policy and research implications

8.1 IMPLICATIONS FOR PRIZE THEORY

The findings of this investigation in four key themes (motivations, R&D activities, technology outputs and overall effect on innovation) (summarized in Table 8.1) have shown that in fact each prize is a unique, complex phenomenon whose understanding requires not only the incorporation of new explanatory factors in our prize theories but also the investigation of other entrant- and context-level factors that affect the development of competitions. Prizes have generally been investigated using formal economic models in which a prize sponsor offers a unique monetary reward (the cash purse) to induce either increasing R&D activity in a specific technological field or the production of a single innovation. This innovation is generally assumed to be placed in the public domain. These models typically involve rational, profit-maximizing R&D performers or prize entrants that factor out monetary benefits, costs and the probability of success in their choices and decisions to participate. Moreover, entrants' pre-existing and/or post-prize activities are not generally considered. This investigation has shown however that the prize phenomenon has strong connections with a number of entrant- and context- level factors, is influenced by multiple both monetary and non-monetary incentives and involves entrants that make decisions based on pre-existing activities, goals that may go beyond the prize challenge, and beliefs and subjective perspectives about the merit and risks of prize participation.

First and foremost, prizes offer a wide range of incentives in addition to the monetary reward. This investigation proposed two analytical categories of incentives involved in prizes. On the one hand there are *prize incentives* that are created with the announcement of prizes and the development of competitions. These include the cash reward and other diverse non-monetary benefits whose perception varies widely according to subjective entrant-level factors. These non-monetary incentives comprise for example opportunities to gain reputation and visibility, participate in technology development and accomplish other personal and organizational goals that include the pursuit of valued ideals such as the

Table 8.1 Summary of research probes, findings and general implications

	Dimensions of prize case study			
	Motivations	R&D activities	Technology outputs	Effect on innovation
Theme	Weight of different types of incentives and relationship with type of entrants	Characteristics of prize R&D activities and difference with traditional industry practices	Prize technology outputs and their relationship with entrants, their R&D activities, and technology incentives	Overall effect of prizes on innovation
Anticipated relationship	H1: Type of entrants depends on types of incentives offered	H2: R&D organization depends on lead times and funding requirements involved in challenge	H3a: Type of technology output varies across types of entrants H3b: Technology incentives induces outputs at higher maturity levels	H4: Innovation effect depends on prize incentives, technology gaps and challenge definition
Dependent variables	Type of entrants (unconventional, conventional)	R&D organization in terms of: Design criteria (simplicity) Technology sources (use of existing technologies) Extent of collaborative effort	Prize technology output in terms of: Degree of novelty of technologies Maturity of technologies	Innovation effect
Independent variables	Type of incentives (prize, technology)	Development lead times/ lack of up-front funding conditions	Type of entrants Technology incentives	Prize incentives Technology gap in prize challenge Openness of challenge definition
Control factors	Technological field Broader context	Technological field	Prize challenge definition	Technological field Broader context

Observed relationship	Prize non-monetary incentives have the greatest incentive power. Technology incentives more likely to attract conventional entrants. Participation is ultimately explained by entrants' goals and perceptions of wide range of incentives.	Designs, technology sources and collaborations are more likely to be associated with individual teams' goals, strategies and resources, and are not directly influenced by prize. Iterative problem-solving activity and convergence of new and ongoing R&D threads is unique to this context.	Some evidence connects type of entrants with prize technologies; other significant factors are prize challenge definitions, current-day technology and ongoing projects. Technology incentives do induce further advances in the innovation pathway.	Innovation effect depends on prize incentives but also on characteristics of entrants, technology sector and broader context; technology gaps induce inventive activity; open-ended challenge definition enables innovation.
General implications	Prizes can selectively incentivize individuals and organizations to advance technologies or pursue related goals using both monetary and non-monetary incentives. Potential markets for the prize technologies complement prize incentives.	Prizes can induce increasing levels of R&D activity and enable the implementation of unorthodox approaches. The development/evolution of competitions however cannot be generally anticipated. Expensive challenges may divert problem-solving efforts.	Prizes can focus on the advancement of technologies at different levels of maturity, but unless challenge definitions are very specific, the quality/characteristics of the technologies are still difficult to anticipate and depend not only on prize challenges but also on context- and entrant-level factors.	Prizes can induce innovation over and above what would have occurred anyway, yet overall effect also depends on context- and entrant-level factors. Technological breakthroughs can be enabled but not directly induced. Technologies may not be use-ready and require further development.

Source: Author's analysis.

contribution of S&T to society or the environment. These incentives can, by definition, be set by prize sponsors although in practice their valuation is problematic from the prize design standpoint. The evidence suggests in this regard that these non-monetary incentives can be overall much more significant than the cash purse. On the other hand there are *technology incentives* linked to the potential market value of the technologies involved in the prize challenge. These incentives play a key role in this kind of prize (particularly in prizes that involve more mature technologies) despite the uncertainty around the costs to accomplish the prize challenge and the actual market value of the technologies.

Because of such a diverse set of incentives, prizes attract entrants with diverse goals. There are entrants motivated by a strong desire to win the competition (*challenge teams*), who thus seek to optimize their efforts to be the first to accomplish the prize challenge (or perform better than anyone else in best-in-class prizes). Their motivations are generally related to the challenging nature of the project or other subjective reasons based in their beliefs about the merit of the pursuit. In practice however most of the prize entrants may be primarily driven by goals other than the achievement of the prize challenge. For example prizes can attract entrants whose priority is starting a new business based on the prize technologies (*startup teams*). These entrants consider the potential market value of such technologies, the benefits of introducing those technologies in own processes, and the added-value of prize participation to their business cases. Their perception of potentially sizable markets for the prize technologies (with less uncertainty about their realization than experts perceive), suggests the presence of overly optimistic cost-benefit calculations or entrants that are less risk-averse than traditional industry players. There may also be another diverse and potentially numerous set of entrants (a *diverse majority*) whose priority is to accomplish other personal or organizational goals and find in prizes the means to do so.

Rather than a profit-maximization problem, at the core of the decisions to participate in these prizes is the belief about the technical feasibility and the social, personal and/or commercial value of the pursuit. The choice between prize participation and the pursuit of alternative paths with similar technological targets depends on either the value the competition adds to entrants' strategies or the opportunity given by the competition to accomplish other goals. Therefore if there are low costs of participation (e.g. small entry fee) prizes are likely to attract more entrants, including those whose goals are not necessarily to win the competition. The benefits of participation are still valuable to them for the pursuit of their strategies. In terms of perceptions of prize incentives this implies that non-monetary incentives can be more important than the monetary reward. Moreover, from this perspective the probability of winning the competition has a diminishing role in the decisions to enter prizes, as late entries exemplify. In other words, in competitions any contender would certainly prefer to win, but in these prizes, even with decreasing probabilities of success, there are still entrants who perceive benefits that result from some kind of good prize performance (for example when entrants seek publicity) or mere participation (for example when entrants seek self-fulfillment).

A probe into the relationship between types of entrant and types of incentive (H1) showed that overall the monetary reward may not be as important as other prize incentives. A sizable cash purse may still be important however to disseminate the idea of the prize. This investigation did not gather specific evidence of this, but the importance of the visibility of competitions suggests that large amounts of cash purse can help to position the competition in the media and attract public attention which is key to showcasing the achievements of entrants, raising interest in the competition. From this perspective the calculation of the monetary reward has to consider not the social value of innovations but the goals of prize programs (e.g. induce increasing R&D, procure technology, promote entrepreneurship, incentivize participation of certain individuals and organizations). The costs of technology development may also be a factor in this calculation if the prize involves technology commercialization and can help to close business cases of entrants (although the prize can also offer other valuable resources to entrants such as access to new professional networks). A sizable reward can also distinguish a competition from other prizes and from alternative strategic paths that entrants might pursue to develop the same kind of technologies.

Prizes can trigger new R&D activities but also induce both the convergence of ongoing R&D processes toward the prize challenge and a significant use of existing knowledge and technical solutions. Prize problem-solving efforts are generally aimed at developing low-cost, simpler solutions but not necessarily at producing solutions of higher quality, performance or market value. Even in cases in which accomplishment and low cost and simplicity are desired attributes, there may still be a potential misuse of resources in the efforts to search for those solutions. Prizes can engage very diverse entrants with distinct interpretations of the problem to be solved and the kind of required solution (and its market value), and hence varying estimates of the cost and risks of achievement. When entrants do not have experience with prize technologies, which is the case of potentially innovative industry outsiders, their estimates and strategies might be poorly thought out and lead to costly trial-and-error iterations. More generally, broader participation of outsiders in prizes can create a competence gap in problem-solving (Dosi and Egidi, 1991) that results from the uncertainty about the means required to achieve the prize challenge and leads to unnecessary expenditures and risks.

On the other hand ongoing industry projects and existing knowledge and technologies contribute solutions to the prize challenge and reduce the costs of overall R&D effort. There is however a potential opportunity cost to attracting efforts toward an ill-defined or not socially desirable prize target and diverting them from more useful purposes from a social or economic standpoint. Moreover, there may still be significant R&D costs that result from adaptation efforts and improvements of existing technologies for use in prize projects. The upside of this process is the potential inducement of user-led innovation when prize entrants successfully adapt or build more efficient versions of pre-existing technologies for use in own projects (von Hippel, 1976, 1977, 1982). Although entrants might not be particularly concerned with other uses of those adapted technologies, previous research found that this kind of user-led innovation may have more commercial merit than regular products when they are generally preferred by other users (von Hippel, 1988).

Prize design parameters such as limited development lead times and lack of up-front funding influence prize R&D activities as they become part of the problem to be solved (Hypothesis H2). But while limited lead times can have the positive energizing effect that makes teams work harder (Amabile et al., 1976) the search for funding diverts problem-solving efforts and might increase the costs of achieving solutions compared to contexts in which research teams have up-front funding (e.g. research grants).

Widespread and collaborative prize participation (with multi-location and multi-organization projects) is appealing to tap into the creativity of many more individuals, leverage efforts and spread the risks of technology development. Widely distributed efforts to find solutions may however be costly and lead to a less efficient problem-solving process from a technical viewpoint, even in the case of more flexible and low-cost organizations that draw on sophisticated communication and virtual collaboration tools. Face-to-face communications are particularly relevant in engineering projects as they enable more rapid feedback, decoding and synthesis of complex information (see for example Kessler and Chakrabarti, 1999). From a broader social perspective there are still benefits to this kind of distributed prize R&D process, such as increasing learning, training and development of solutions that ultimately might be useful for other projects.

Since prizes are generally open and widely publicized public processes, they increase opportunities for knowledge exchanges and the implementation of open innovation approaches that leverage external research and complement internal technological activities with increasing knowledge flows (Chesbrough, 2003, 2006). Prizes also offer opportunities for public engagement and crowdsourcing, or outsourcing of tasks to the general public (Brabham, 2008; Kleemann et al., 2008). For example, although none of the case studies of this investigation was designed specifically for this, a few GLXP entrants report using ideas and comments received through their websites to inform their R&D processes. This kind of broader collaboration in the context of the competition might also serve as a basis for the formation of stronger problem-solving communities that transcend the prize time frame (Bullinger et al., 2010; Hutter et al., 2011) but at the cost for prize sponsors of promoting further communication and collaboration.

Although sponsors do not face a moral hazard problem or R&D risks involved in the development of the prize technologies, they still face great uncertainty about the evolution of prize competitions (and hence the costs of implementation) and social costs of searching for solutions to the prize challenge. The sources of this uncertainty include the difficulty of anticipating R&D approaches, uncertain cost functions of R&D performers due to the diverse forms of R&D organization, and the difficulty of measuring the extent of actual participation that includes not only entrants but also networks of partners and collaborators. Hence the lack of information on the costs of finding a solution, in order to offer the appropriate incentives and set other prize design parameters, is potentially much more problematic and involves more factors than the previous literature suggests. Moreover, in the course of competitions sponsors may still need to monitor entrants' activities to learn about constraints in R&D processes and guarantee a positive evolution of the competition.

The uncertainty involved in the evolution of prize competitions also exists in relation to the technology outputs of prizes, despite a number of prize design parameters available to prize sponsors. Prize technologies also build on industry projects and readily available technologies, and depend on entrant-level factors such as goals, skills and resources. Moreover, the wide range of prize technologies generally addresses, rather than the production of a single innovation, an entire solutions space. An analytical concept like this is necessary to describe such a range of technologies in terms of key dimensions. In the case of the GLXP for example the solutions space can refer to mission approaches and comprises missions with budgets between $5 million and $100 million, with development lead times between 30 and 90 months and with funding from varied sources. Prize design parameters can influence how focused this solutions space is, but not the quality, commercial merit or other attributes of the approaches. The performance of the technologies can only be influenced if the prize challenge is defined as a measure of performance (i.e. best-in-class prizes) in which case the solutions can be expected to converge to such a measure or its improvement.

From this perspective the effectiveness of first-to-achieve prizes depends on the balance of sponsor interests and the uncertainty resulting from unanticipated characteristics of the prize technologies. Hypothesis H3 sought to shed light in this regard and probe the notion of unconventional entrants that contribute fresh ideas and unorthodox approaches, and the influence of technology incentives that complement prize incentives. At the entrant-level the goals of entrants are a more important predictor of the technologies they work on. Entrants generally optimize their performance to reduce R&D efforts and draw upon simpler and current-day technologies in their attempts to win competitions. Entrants with other goals pursue a range of strategies and may produce technologies of diverse characteristics, but these cannot be directly influenced by prize design. For example, when there are significant technology incentives, prizes can induce the advancement of technologies at higher maturity levels for commercialization even if the prize does not require actual commercial activity for the reward to be claimed. Prize sponsors can set stricter technical specifications to further focus technology development but at the risk of over-focusing the prize and possibly constraining the incentive effect and innovative activities induced by the prize. Too-detailed technical specifications might impede prize sponsors efforts to reward other valuable innovations resulting from the process. Open-ended challenge definitions and relaxed technical requirements can enable innovative solutions but at the cost of broadening the ultimate solutions space and potentially leading to solutions that require costly adaptation or development efforts to meet prize sponsor's specific needs. Moreover, these outputs might include technologies that effectively achieve the prize challenge but are still potentially ill-designed, inefficient, low quality or risky.

The findings of prize cases with technological goals at different points in the creation-to-implementation innovation pathway (GLXP at higher TRLs, NGLLC at medium-high TRLs and AXP at medium TRLs) also suggest that prizes may not be equally effective for all technological goals in a given field. Due to the sequential and cumulative nature of innovation whereby each successive invention along the innovation pathway builds in an essential way on its predecessors (Green and Scotchmer, 1995), relevant knowledge and technologies have to be available to competitors for prizes to successfully induce technology research, development, improvement or diffusion/commercialization. In other words the larger the tool-kit readily available to address the problem, the higher the technology maturity levels the prize program can aim for. Otherwise, entrants would need further incentives, support and/or longer development lead times to be able to achieve the prize target. More generally, prizes (a) may not be effective to induce technology development if they pose challenges for which relevant knowledge and technologies are not yet up to the prize requirements, or (b) may effectively produce such effects but only if they allow longer development lead times and provide more significant incentives and/ or development support. For example, prizes might not be effective to induce commercialization of technologies if these technologies are still in early proof-of-concept phase. Or, when sponsors seek to induce basic research or discovery of new principles in unexplored fields, longer-term prizes might be required, to increase experimentation and testing. This perspective calls into question whether prizes have the ability to induce innovations that require expensive or very long-term R&D efforts, as suggested in a number of proposals by scholars and prize enthusiasts to seek solutions to critical social and economic problems.

The generalization of this perspective is also the basis for a typology of prizes that can become a source for hypothesis testing and case study modeling in future research (Table 8.2). This typology is primarily characterized by a relationship between the prize target and current-day capabilities and technologies readily available in the sector in which the prize is announced. The more mature the technological solution the prize aims for, the more knowledge background and pre-existing technologies are required for entrants to be able to produce a solution within given development lead times and potential costs of achievement. This relationship yields distinct generic applications of prizes which involve different kinds of technological gaps. Prizes that focus on lower maturity technologies or the discovery of new scientific principles have the highest programmatic risk and are likely to lack complementary technology incentives to help entrants to raise funding for their projects or to push technologies forward to implementation/commercialization. On the other hand prizes that involve technology implementation/commercialization are more likely to depend on favorable conditions of the context and the existence of actual markets for the prize technologies. These prizes certainly require key relevant technologies that are readily available, unless longer lead times are allowed for problem-solving. Prizes for technology development and incremental improvement are at the mid-point of the innovation pathway and also require either knowledge of basic principles and experimental research already performed or technologies with medium-levels of maturity, respectively.

In a context in which corporate and academic R&D choices are generally limited to other traditional incentive mechanisms, prizes can induce innovation over and above what would have occurred anyway. Technology advancement in prizes results from new R&D efforts and the

Table 8.2 Typology of prizes according to targets along the innovation pathway

	Types of prizes			
	Prizes for discovery and novel solutions	Prizes for technology development	Prizes for incremental improvements	Prizes for technology implementation
Technology gap that entrants have to close	Nature of solution unknown	Technical feasibility not yet demonstrated	Feasibility demonstrated but technology still not efficient	Technology feasible and efficient, yet not implemented/disseminated
Generic challenge definition	Discover or create a method to perform new function: 'function to be performed'	Develop technology to accomplish a feat: 'feat never achieved before'	Improve technology to achieve higher performance standards: 'improved performance target'	Implement technologies under prize challenge timing/cost conditions: 'repeat achievement under new conditions'
Current-day technology available to entrants	Unclear what basic principles the solution should draw upon	Basic principles known, experimental research performed	Functional technology with at least medium level of maturity	Key technologies with medium-to-high level of maturity
Generic applications of this kind of prizes	High-risk, exploratory approaches to find 'out-of-the-box' solutions	Demonstrate technological feasibility	Advance technologies with specific (commercial or other) applications	Accelerate technology diffusion, adoption or development of end-user communities or markets that are held back
Potential prize examples	Early prizes? (e.g. Food Preservation Prize, Longitude Prize)	AXP	NGLLC, DARPA Challenges	GLXP

Source: Author's analysis.

continuation of R&D efforts that are held back for diverse reasons. There is user-led innovation as entrants develop technologies for their own prize projects and there is advancement of technologies for commercialization or implementation in other projects. There is no evidence however of prize factors that directly induce breakthrough innovations. The ability of prizes to induce innovation is larger when there are larger prize incentives, more significant technology gaps implicit in the prize challenge and open-ended challenge definitions (H4) but there are other intervening factors as well. In simple terms larger prize incentives (both monetary and non-monetary) can attract more, and more diverse, entrants that undertake R&D efforts and potentially produce innovations (which may or may not directly address the prize challenge). The incentives elasticity of innovation however is likely to vary across fields and depends on a number of entrant- and context-level factors (such as the pool of skills the prize attracts and the conditions for fundraising, respectively). Increasing technology gaps pose more significant technical problems to solve and hence increasing opportunities to come up with innovative solutions. Open-ended challenge definitions offer rewards and opportunities for the advancement of a wider range of technologies.

In a context in which there is a widespread use of prizes their effects might also depend on the uniqueness of competitions in terms of prize target and sponsors. More generally, a prize announced in a context in which no other equivalent competition (or any competition) is simultaneously held is likely to have more incentive effect than a similar prize in a context in which other rival prizes also seek to attract entrants, resources and the public's attention. In practice the routine use of prizes and challenge definitions that overlap might weaken the incentive power of the mechanism. The AXP, NGLLC and GLXP have not had rival prizes held simultaneously and therefore do not fully represent that situation. Moreover, in the process of disseminating the idea of the prize and the achievements of entrants, media attention to specific competitions may fade if an increasing number of prizes are announced.

The extent to which the prize mechanism depends on its context and the fact that in principle winning a prize is not a business sustainable over time suggest that prizes are not able to create markets by themselves. On the one hand the conditions of the general context can frustrate (or facilitate) technology sourcing and fundraising efforts of entrants. This is more relevant for prizes that involve expensive R&D projects and for prizes aimed at commercialization of technologies, whose success also depends on the perceptions of entrants and their financiers about the market value of the technologies. On the other hand prize rewards generally represent one-time revenues for the winner (unless there is a commitment of the sponsor to acquire the technology) and therefore may not help to generate a stream of revenue for entrants that seek to create a commercial enterprise (there may exist, although evidence of this is not strong, an emerging generation of serial-prize entrants that seek to profit from multiple prize participation). Sponsors can systematically set rewards that are lower than expected R&D costs if prizes offer other valuable opportunities, for example to increase reputation and visibility, which are desirable factors for new start-ups in certain industry sectors. From this perspective prizes can support new R&D performers that seek to enter more mature markets if they actually succeed in finding better technological solutions to serve markets related with the prize technologies.

The case studies also showed that prizes complement and do not replace other incentive mechanisms available to entrants, and help to better explain the ability of prizes to leverage R&D funding. From the point of view of R&D performers, the choices between prizes and alternative technology development paths through other incentives (such as contracts, grants and patents) are not necessarily rational profit-maximizing decisions. They depend greatly on subjective perceptions of the value added by non-monetary benefits of prize participation to individual or organizational strategies, and are not exclusive in the sense that they are made in a broader context in which performers pursue other economic opportunities and diverse personal and altruistic goals. More specifically modern technology prizes (which systematically offer rewards equal to or below expected R&D costs) complement and do not replace patents and other incentive mechanisms. It is actually the ability of entrants to draw on other incentive mechanisms that helps to attract funding and in-kind resources from industry and external sources, particularly when the technology and funding gaps implicit in the prize challenge are significant and the prize is aimed at technology diffusion or commercialization. For example entrants trade IP rights as a means to get access to key technologies and other resources such as expertise and labor for their projects. In other words IP rights ease collaborations in prizes. Patented prize technologies are then disclosed but not in the public domain as theoretical models generally assume. Entrants also engage in contracting opportunities and seek funding through alternative means which may include grants as well.

Prizes might lead to a socially undesirable duplication of R&D efforts in the sense that there may be many entrants in the pursuit of the same technological goal when prize challenges are narrowly defined (Maurer and Scotchmer, 2004; Newell and Wilson, 2005). The evidence suggests however the development of very diverse technologies (particularly at the subsystem, part and component levels) and very unique efforts to solve prize challenges (and many intermediate technical problems) even in competitions with larger numbers of entrants. Prize R&D activities are generally developed in threads that represent work to advance technologies at different maturity levels and involve a range of efforts from adaptation of current-day technologies, to incremental improvements, to the creation of novel technologies. Each of those threads addresses very specific technical problems that are approached from diverse perspectives. Any duplication of efforts in this context might be beneficial and a reason to implement prizes. Even if entrants work on similar technologies to achieve the prize challenge, the side-by-side comparison of these technologies at the subsystem- or lower levels and the overall R&D and project management approaches are a valuable source of knowledge for the sponsor, entrants and industry players. That kind of comparison however requires further disclosure of information on projects and technologies involved in the prize. There are on the other hand social and economic benefits to increasing numbers of individuals and organizations involved in competitions, collaborations and knowledge spillovers, which favor technical training, education, increasing interest in S&T and the achievement of other personal and organizational goals of participants.

8.2 R&D PROGRAM AND POLICY IMPLICATIONS

R&D program managers in public and private organizations can use prizes for multiple purposes. Properly designed prizes can incentivize increasing R&D along different points of the innovation pathway, from pure research to commercialization/diffusion of technologies, but also induce other related outcomes. Appropriate prize challenge definitions can set targets on specific technologies and a range of eligibility criteria for prize participation can help to target specific communities of R&D performers and geographic areas as well. More specific potential effects of prizes may include new forms of collaborative R&D efforts, participation of both traditional industry players and outsiders, and engagement of communities of interest (e.g. students, women, minorities or 'lone garage inventors'). This wide open participation helps to democratize innovation processes and offers the opportunity to tap into more widely distributed creativity and fresh ideas. Prizes can leverage programs' R&D funding significantly due to their widespread, decentralized impact but at the cost of higher programmatic risks than other more traditional instruments. Prize R&D activities source ideas broadly by their own nature and can induce the implementation of new, more flexible, efficient and/or collaborative approaches to problem solving if they do not pose restrictions to R&D organization, overcoming the limitations of other more traditional instruments.[1] Broad prize participation also suggests an opportunity to compare R&D approaches and technologies side-by-side which, regardless of the prize outcome, is already a valuable source of knowledge to inform future R&D decision-making. Prizes might not be able to create new markets but their ability to induce the continuation, acceleration or re-direction of existing projects suggests potential uses to sustain R&D investment to address certain targets. Prize competitions can also showcase the technologies and effort of participants and influence industry decisions in this way.

Prizes may induce other valuable outcomes that are more difficult to anticipate than the immediate effects of prize announcements. Prize R&D activities may generate significant knowledge spillovers when they bring together new and pre-existing small and large companies, universities and NGOs to develop prize technologies. Collaborations and other relationships developed in competitions might transcend the competition time frame and create problem-solving communities. The kind of iterative problem-solving efforts induced by prizes are unique opportunities to learn by doing, gain hands-on experience and train individuals. Increasing visibility of prize competitions may also influence industry and public perceptions, raise public awareness and change beliefs about S&T topics linked to the sponsor's mission. Successful prizes, both historic and modern, have not only attracted individuals that believed in the feasibility and merit of the prize goal, but have also contributed to spread that belief to industry and the broader public. Success stories such as the Orteig Prize are the foundation of modern prizes and have inspired entrepreneurs, philanthropists and other individuals that seek opportunities to participate in challenging projects. Sponsors of modern prizes have more powerful communication means at hand to disseminate the achievement of competitors and transform beliefs about scientific and technological possibilities. In some cases the inspirational value that prizes create for innovations to come might be as important as the immediate effects of their implementation.

Prize program managers can take advantage of the number of prize design parameters and the range of potential effects to achieve their organizational goals, but the prize process still involves significant uncertainty and programmatic risks. Most importantly, it is difficult to anticipate the evolution of prize competitions and their technology outputs. Later outcomes are even more difficult to predict. This raises concerns about the efficiency of the R&D program spending from budgetary and societal standpoints. Prize managers can anticipate neither the approaches that entrants use to find the solution nor the levels of R&D activity. Moreover, if prizes involve challenges that are very expensive to achieve, entrants may divert efforts from R&D considerably to seek funding, and technological development may be constrained. Hybrid prize schemes that include financial support for qualified entrants (such as R&D grants) or commitment to purchase prize technologies (such as procurement contracts) are potentially more efficient designs for competitions that involve ambitious targets. Prizes can also induce a more focused R&D effort by providing seed funding or other forms of in-kind support such as access to special facilities or equipment for all entrants. Unfortunately, the very uncertainty that prize R&D processes involve impedes accurate calculation of the R&D costs and thus the optimal level of support. Prize sponsors may still want to induce fundraising efforts and attract funding from other sectors to incentivize the development of commercially viable solutions.

Prize R&D efforts may also be inefficient and originate activity that ultimately does not result in, for example, the introduction of the most efficient and safest technologies or the dissemination of the most appropriate method to address a problem. Uncertain methods of technology development in the pursuit of certain technological goals may also involve higher risks for individuals, the environment and property. Therefore, program managers have to implement the necessary measures to limit liabilities and use prizes only as an experimental kind of program when there is high uncertainty on the potential outcomes of competitions. Moreover, although the introduction of unorthodox approaches is an appealing idea in some circumstances, further work may be necessary to properly introduce and disseminate those methods as standard practices. For example codification and documentation of procedures and methods may be necessary if entrants are informal organizations that rely heavily upon trial-and-error or other informal approaches to R&D.

This uncertainty involved in the prize process calls into question the efficiency of prize program spending and suggests that prizes should be used as a component of a broader portfolio of actions and instruments to incentivize R&D and induce innovation, to spread the risks involved with prize implementation. Interactions with other instruments can also create synergies and increase the effectiveness of prize programs: for example when entrants require some kind of support for R&D activities and other programs, the prize sponsor can offer research grants or seed-funding for prize projects. To attenuate programmatic risks, prize sponsors can also partner with other entities to organize competitions. There may be a limit to the widespread and routine use of prizes if challenge definitions overlap or some prizes attract most of the attention. To avoid this, program managers may also use a portfolio perspective. For example, prizes can initiate new lines of research and contracts to support further development of prize outputs. If a multi-prize program is designed, program managers can implement sequential competitions that build on previous results and posit increasingly difficult challenges, simultaneous yet complementary global competitions, or similar yet regionally focused competitions.

The choice of prizes from a cost-benefit perspective (in prizes v. other traditional instruments and between alternative prize designs) has to consider that the monetary costs of prize programs may significantly exceed the amount of the cash purse. This is particularly relevant if sponsors seek to advance specific technologies but engage large numbers of entrants that may or may not target the prize challenge, which would increase administrative costs without focusing efforts on the actual prize problem. These costs are distributed among the design phase (e.g. expert surveys, design workshops), prize launch and administration (e.g. promotional actions, survey of participants and outputs), actual reward (if the prize finds a winner) and post-prize related activities (if for example technologies require adaptation/further testing before they are ready to use). Prize sponsors also have to consider that there are diverse measures of direct and indirect benefits in prize programs, as the range of immediate and later outcomes is diverse as well. Sometimes the value of later outcomes may be more important than the market value of the technologies from the point of view of the sponsor (when for example the prize sponsor is interested in the dissemination of ideas or the engagement of individuals for training). In this context the risks include those related with R&D failure and technological uncertainty, the programmatic risks that emerge from program failure/ill-designed prizes, and the more general risks for health, property and the environment.

R&D program managers also have to consider that the implementation of prizes might be efficient only in certain technological fields. Prizes have to pose targets for which the relevant knowledge and technologies are readily available to entrants. Timing is key in this regard. If prizes target less mature, emerging technologies or unexplored technological fields they have to allow increasing development lead times and possibly provide some kind of monetary (e.g. seed funding) or in-kind support to help entrants to maintain and not divert their R&D efforts. This is because prizes may induce the introduction of novel concepts and ideas that are in early stages of the innovation pathway but still far from actual implementation or use. Appropriate timing and funding conditions in prizes however can still induce the implementation of potentially low-overhead and more efficient project management approaches and the development of new funding sources and revenue models.

The prize design phase is very important to assess industry capabilities in the field and provide the right incentives according to specific program goals. But there are other contextual factors that can influence prize program success rates. Broader economic conditions may facilitate or moderate fundraising when prizes require more significant R&D efforts and marketing when prizes require commercialization of technologies. On the other hand existing legal frameworks and regulations may impede technology development, testing and deployment. In both AXP and NGLLC for example, only FAA's space flight and experimental permits allowed teams to fly their vehicles. Since prizes may induce collaborative efforts across geographical boundaries, rules have to consider the regulatory framework specific to the industry sector, particularly when prizes involve technologies that may have dual-use or be considered inherently military in nature. Other special regulations might be required to protect health, property and the environment in the development of competitions.

The case studies presented in this book contribute insights on aspects of prize design, implementation and evaluation. The following insights refer to key parameters of prize program design:

- *Prize challenge*
  Prizes should address concrete problems that represent a significant technology gap, but not necessarily require enormous R&D expenditures (e.g.'establishing the first human moon base'), and whose achievement is likely to be clear and verifiable. For this the challenge has to be clearly enunciated and easy to communicate. Moreover, achievements that are visible to the naked eye are much more attractive to the media and the public (who would be able to follow the activities and attempts of entrants) and support the selection of fair prize recipients. Prize challenges should be attractive and captivating for a certain number of potential entrants that believe the challenge is feasible and has some social, personal or commercial merit. Prizes that target 'good causes' or are related with other widely shared beliefs are likely to have wider participation. Conversely if there is a conflict between the spirit of the prize and the points of view of entrants interest in participation and excitement might vanish. Challenges aligned with sizable markets on the other hand can bring in additional incentives to further advance technologies toward implementation and commercialization. The definition of the challenge includes setting a prize deadline that allows, according to expert opinion (that is, according to entrepreneurs, industry leaders and scientific committees for example), a reasonable lead time for technology development and time for team formation and growth. Depending on the ultimate goal of the prize program, the challenge definition can be more or less open to allow the introduction of innovative approaches and creative solutions or induce the

advancement of specific technologies. Open-ended challenge definitions offer more, and more diverse, opportunities to entrants. These
are opportunities for prize sponsors as well, for example to operate
competitions as a test bed for unorthodox methods and radical technologies when the prize challenge is defined in terms of achievement
with no requirements to build specific technologies.

- *Cash purse and other prize incentives*
  Competitions should balance the prize purse with the potential
  market value of the prize technologies and other benefits that the
  competition may offer to participants. The prize reward has an
  important role in distinguishing the competition from other prizes
  and economic opportunities. It should involve an amount that
  attracts the attention of the media and general public. The cash
  purse is also a potential support to start new businesses. The evidence from these case studies suggests however that a reduced cash
  reward and a small amount of up-front funding offered as seed
  funding for projects is more likely to stimulate the problem-solving
  process than a larger monetary reward with no up-front funding
  and only a small probability of victory. Prizes can also create
  attractive non-monetary incentives. A challenging project can drive
  curiosity and the desire to participate and compete. Prizes can also
  offer the opportunity to gain reputation and publicity, and create a
  competitive environment that inspires people that seek recognition.
  Low entry-barriers to the competition also incentivize participation
  and give access to a set of resources available only through the prize
  competition. In particular the reputation/publicity value created by
  prizes supports the strategies of entrants that seek to enter markets
  that require a proven track record in successfully delivering technical solutions, such as aerospace, defense and medicine. The official
  endorsement of the competition aimed at technology commer-
  cialization/diffusion by key industry players or industry/consumer
  organizations can increase the incentive power of the competition.
- *Who is eligible to participate*
  Prizes can also target individuals and groups of diverse age, professional background or experience by defining special criteria of eligibility or offering particular incentives. Target communities may be
  defined for example as engineering students, teachers or government
  employees. This makes prizes an appropriate mechanism to engage
  people in science, technology, engineering and mathematics (STEM)
  education or technical training programs. The eligibility criteria can
  be relaxed to attract more, and more diverse, entrants, but this can
  increase the cost of program operation significantly and change

the spirit of the competition. Online prize platforms created for competitions are another means to increase other forms of indirect participation and collaboration. More generally prizes may effectively attract individuals and companies that are less risk-averse for both technical and commercial endeavors. Outsiders are interested in challenging projects. In particular prize challenges that balance a sizable market potential with uncertainty about the market segments, market values and required capabilities to exploit them are more likely to engage less risk-averse entrepreneurs and discourage traditional industry players. Prizes can also be implemented within specific regions or seek broader participation to tap into widely distributed ideas and creativity.[2] At the state/city level, prizes may help to mobilize resources into underserved areas. Prizes should balance the openness of the competition and the eligibility criteria (through the implementation of fees or a procedure to evaluate entry applications) to tap into widely dispersed creativity but allow only serious entries. Special eligibility requirements might be needed when industry-specific regulations demand them.

- *Prize rules*
  The evidence highlights the importance of simple, unambiguous, transparent and easy to understand rules. They have to be fair to create a truly level playing field. They have to remain unchanged after the competition has been announced to avoid discouraging participating teams. The set of rules can be significantly improved if experts and potentially interested participants contribute insights in this process. Design-wise prizes are very flexible and allow setting multiple parameters, for example to test special regulatory frameworks and compare competing approaches to R&D, technologies and business strategies in a level playing field.

Selected examples of prize designs help to illustrate how specific parameters can be set to address different goals (Table 8.3). These examples include prizes aimed at: (a) exploring new, experimental methods and technologies that imply high-risk R&D; (b) inducing technological development to break critical technological barriers; (c) accelerating technological development to achieve higher performance standards; and (d) accelerating the commercialization of technologies.

The analysis of industrial and economic trends can help to anticipate favorable contexts and the potential influence of external factors when competitions are launched. Prize announcements can take advantage of related public events and strategically increase the prize program's visibility, reaching out not only to those that may eventually enter the

Table 8.3   Examples of prize-based program goals and definition of key prize design parameters

	Program goals			
	Explore new, experimental methods and technologies that imply high-risk R&D	Induce technological development to break critical technological barriers	Accelerate technological development to achieve higher performance standards	Accelerate commercialization of technologies
Type of prize	Technology-based accomplishment	Technology development (demonstration of technical feasibility)	Technology development (demonstration of increasing performance)	Technology-based accomplishment
Expected outputs for prize sponsor	Experimental technology, knowledge of comparative features of new R&D methods	Feasible technological solutions that can be further improved	More efficient technology	Increasing commercial activity in technology markets, new technology and service providers, new business approaches
	Design parameters			
Prize challenge definition	Discover or create a method to perform new function	Develop new artifact/system to accomplish a feat (newly defined problem)	New performance requirements for existing technologies	More challenging timing/cost conditions to accomplish a feat linked to specific commercial markets
Monetary reward	Amount that is attractive for independent inventors, professionals	Amount that is attractive for smaller industry players as startup funding	Amount that covers part of costs of incremental development	Amount that can potentially close business cases

Non-monetary incentives	Endorsement of scientific and professional societies to offer prestige and potentially funding for further research	Opportunities to obtain funding for technology development	Endorsement of potential customers or industry leaders	Endorsement of potential customers or industry leaders
Market value of technologies	There may not yet be markets for emerging technologies	Prize challenge may be linked to sizable yet uncertain markets	Sponsor may offer commitment to license or purchase technology	Prize challenge linked to markets that are held back for diverse reasons
Additional considerations	Entrants might need R&D support in the form of seed funding or grants; may be implemented as prizes that build on previous prizes to attenuate programmatic risks	Entrants might need R&D support in the form of seed funding or grants	Set concrete, verifiable measures of improvement to find a fair winner	Potential markets required; entrants must retain IP

Note: Other more general considerations apply as described in the rest of this chapter.

Source: Author's analysis.

competition, but also to broader audiences including industry officials, policy makers and the general public. The process of communicating and disseminating the idea of the prize is particularly important for prizes that aim for global scale. Online social network platforms and other communication means can help in that regard and engage more individuals with diverse backgrounds and experience and thus enable a more innovative technology development process. Longer registration periods allow broader dissemination of the idea of the prize and more time to make the decision to enter. Increasing activity in the competition during this initial period can attract the attention of more participants.

Other kinds of prizes with technological targets may not have similar effects, dynamics and interactions with their contexts. These three prizes posited ambitious challenges, offered notable cash purses and were linked to sizable potential markets in a favorable period of transition of the space sector toward private investments and a more commercial orientation. Less ambitious challenges are less dependent on the conditions of the context for investment. Smaller cash purses can still be attractive if prizes focus on particular regions or communities. Also, prizes with education and participation goals can facilitate wider participation. Those prizes aimed at promoting science and technology and engineering awareness and careers (e.g. Quest Challenges in Davidian, 2005) and prizes aimed at strengthening communities (e.g. Network Prizes in McKinsey & Company, 2009) for example may effectively stimulate individuals with smaller cash purses, less ambitious targets and eligibility criteria that favor diversity and encourage participation.

After the prize has been launched the sponsor, or the administrator chosen for the competition, must continually assess the activities of the participants and the feedback provided by them during the execution of the program in order to anticipate potential problems and maintain an exciting competition with the engagement of the media and the public. When the public can watch and follow a prize, it is inherently more exciting for both the participants and the spectators. To promote the competition and the activities of entrants, prize sponsors may appoint media relations managers and create websites with information about the prize, news releases and profiles of the participant teams. This is particularly important in competitions that are not held at a pre-specified site or do not include a competition day, such as the AXP. In competitions like the NGLLC a more open format allows for greater public involvement. The final events of those competitions were held at a pre-specified site, were open to the public and attracted thousands of people interested in seeing the teams compete.

Depending on the goal of the prize program, sponsors may seek to use the technologies developed in prizes. The key role that IP rights play in fundraising and commercialization activities of entrants puts forward the question of whether entrants should be allowed to retain IP rights when program managers want to further disseminate or advance technologies with follow-up competitions. If prize sponsors are willing to further disseminate or advance prize technologies, they should enter in agreements to negotiate in good faith to license the technologies or have preferential access to commercial services offered by entrants.

Prize program evaluation requires special considerations compared to other incentive mechanisms. The generally widely distributed R&D effort in prizes is difficult to measure and therefore the assessment of the overall impact of the program is challenging as well. The effect of prizes can be observed from the very announcement until long after the competition has ended. But the ultimate outcomes of a prize program, including those that come unexpectedly, may not be observable until months or years later. The implications for prize program evaluation are twofold: first, multiple and diverse metrics should be used to evaluate programs; and second, there should be multiple evaluation points in the program timeline, which will vary depending on the program. In developing metrics, program managers need to define the appropriate time horizon for measurement based on the characteristics of each prize program. In developing metrics for investment leverage, program managers should consider that these are typically fuzzy and difficult-to-measure concepts in the context of prizes. Prizes may also induce more R&D activity by organizations not officially registered for the competition in the form of follow-up investment and post-prize achievement of prize entrants (e.g. new contracts, new funding). Particularly in government prizes, surveying prize entrants and partners to gather data on their activities may raise confidentiality concerns on their part and even have a negative effect on the ability of the prize to attract entrants. Program evaluations also have to consider the dynamics of the competition in terms of entrants (new teams, drop outs, mergers) and significant variations in the number of members/volunteers that teams engage. Data gathering to evaluate programs is likely to depend on self-reporting instruments and thus require appropriate design to favor data reliability and avoid unnecessary bureaucracy.

8.3   METHODOLOGICAL CONSIDERATIONS

This research has investigated prizes and the means by which they induce innovation using an empirical, multiple case-study methodology and multiple types of data sources. The investigation was set out to answer four questions and probe four corresponding hypotheses that are deemed relevant from the point of view of both scientific inquiry and policy making. Based on insights from previous research, this iterative research process has introduced an innovation model applied to prizes, tested and revised the model with the analysis of pilot case studies, investigated the main case study, probed hypotheses, revised theoretical aspects and drawn implications.

The analysis has been able to provide some answers to research questions and contribute knowledge for a better understanding of innovation prizes. The hypotheses, which reflect assumptions implicit in the prize literature and incorporate other factors intuitively, helped to frame a systematic data collection process for prize cases and triggered a more general yet fundamental discussion about the real effect of prizes. The probes led to the conclusion that prizes do induce technological innovation under certain conditions (H4,) but the underlying dynamics of this phenomenon (implicit in H1, H2 and H3) is more complex than generally assumed. The analysis has unveiled other intervening factors than those hypothetically anticipated and allowed setting a basis for further investigation and refinement of our approaches to investigate prizes. Future investigations should develop hypotheses that incorporate those factors and probe for example the power of certain incentives considered individually, the entrant-level determinants of R&D configurations and the relationship between more specific technical characteristics of prize outputs and prize configurations.

This analysis of the effects of these space prizes on innovation considers the peculiar characteristics of space technologies. Space projects are discrete, one-off products that, generally, to be deemed truly innovative, must satisfy pre-established requirements and meet specific performance criteria or properties (Bain et al., 2001). Moreover, the main players (and potential technology customers) in this sector are very limited in number and generally are those that establish the project requirements and develop the roadmaps for future technology development. The investigations of the effect of prizes linked to, for example, consumer product technologies may differ substantially from this analysis, as innovation in those other sectors is driven by information that companies have on markets with numerous customers and non-obvious inventions companies introduce to address them.

The case-study-based iterative approach is demonstrated to be appropriate to investigate prizes when there is a lack of prior research. The test cases helped to refine the model and provide methodological insights to investigate the main case. The main case contributed significant empirical evidence to gain better understanding of the phenomenon. Overall the three cases contributed insights to develop new building blocks for a theory of prizes. Further research however should investigate and compare two or more cases and cases in different technological fields and broader contexts, rather than following that sequential approach. That will help to gain better understanding of the potential of the prize mechanism under more diverse circumstances.

The innovation model proposed by this research has some advantages over the traditional economic modeling of prize mechanisms. Other models have generally considered only the monetary incentive and the production of one innovation and have not considered other diverse motivations, the indirect participation of for example partners and volunteers and the development of technologies that do not necessarily focus on the prize target. In this regard the most important contribution of this research is an alternative model of innovation in prizes that do consider those factors and in particular assumes that R&D performers are not necessarily profit-maximizing and have diverse decision-making processes, knowledge and skills. Moreover, this new model allows comparison of modern and early prize cases to understand the influence of the industry sector and broader context on competitions, and comparison of prizes with other instances of innovation. The investigation of the prize-, context- and entrant-levels is important and enhances our understanding of the phenomenon. Improvements in this innovation model applied to prizes should include the refinement of the operationalization of research categories, the specification of other relationships between categories and the investigation of new themes emerging from them.

The investigation of prizes without much prior empirical research required probing classifications of incentives and entrants to allow the operationalization of certain constructs. The classification of incentives into prize and technology incentives was appropriate for one of the first empirical research projects that investigates modern prizes. This research has shown how different types of incentives weigh on the decision of entrants to participate but has also highlighted the need to further address the topic in future research to assess the importance of each individual component of the set of incentives. On the other hand this research has shown that entrants are very diverse and that new classifications can be explored. The classification into conventional and unconventional is costly from an analytical standpoint, because it does not allow understanding all the diversity and complexity of entrants. Considering the implementation of prizes, it is still useful for prize sponsors to know generally whether they will be able to engage individuals and organizations not familiar with the prize technologies. Yet the priority goal-based classification (i.e. whether the entrant seeks to win the competition, start a new business or other) may be more illuminating to understand strategies and the factors that can help to focus R&D efforts, despite its more complex operationalization.

There is still little empirical evidence on prize cases. The data available on recent competitions have not been systematically collected and are mostly contributed by anecdotal accounts and media coverage. Hence this research sought to draw upon multiple types of data sources and data triangulation to increase its internal reliability. Visits and observation of team activities in particular have helped to gain a better understanding of entrants' organization and strategies and also yielded valuable insights to better interpret data from questionnaires and documentary sources. This shows that real-time data from ongoing competitions can also be more insightful than historical accounts to understand certain aspects of prizes.

Future research should seek to develop, among others, methods to quantify and qualify the magnitude of the collaborative effort that includes volunteers and partners. For this researchers will likely draw upon self-reporting methods to gather data about entrants and their activities and collaborators, using questionnaires that balance more, and more detailed, questions with simplicity and low bureaucracy to increase response rates and consider the potential issues that self-reporting methods might introduce into the analysis. Potential data sources for future research also include the data generally collected by prize sponsors about the activities of teams. This project for example has had the collaboration of the XPF for data gathering. Future projects should explore methods to gather data systematically without interfering with competitions. When documentary and other third-party data sources are used, triangulation becomes very important to remove inconsistencies and increase the internal reliability of investigations, which is particularly relevant in the investigation of early prizes.

Finally this research has also shown that the winning entry is not necessarily the most important technological development in the context of prizes. Findings of significant activity during the competition (in both teams and their partners) supports the idea that the most interesting effect of prizes is not only the winning entry but also the activities and outputs of runners-up, other entrants and other individuals and organizations that participate in competitions only indirectly.

NOTES

1. A comparison with the US Small Business Innovation Research (SBIR) Program and the Small Business Technology Transfer (STTR) Program is illustrative in this regard. SBIR/STTR awards are competitive with expert assessment. Between 20 and 25 per cent

of proposals are awarded. The award establishes a contractual relationship between the awardees and the sponsor agency (NASA, 2012). This kind of instrument however does not allow the type of collaborative and networked development observed in prizes as they require in-house technology development execution and no partnership. Moreover, while the SBIR/STTR program funds certain R&D effort, specific technology maturity levels, and standard development lead times (12–24 months), prizes allow setting these parameters discretionarily, based on the sponsor's needs.

2. For the aerospace industry for example the GLXP has represented the opportunity for certain countries to engage in the development of space technologies through the participation of individuals and private organizations, an opportunity that probably they would not have had otherwise.

9. Conclusions

Inducement prizes, where cash rewards are given to motivate the attainment of targets, have long been used to stimulate individuals, groups and communities to accomplish diverse types of goals. Lately, prizes that reward the achievement of technological targets have increasingly attracted attention due to their potential to induce path-breaking innovations and accomplish related goals, such as economic recovery or the engagement of social groups to create innovation communities. Governments have become more and more interested in these prizes and, particularly in the USA, sought to include this incentive mechanism within the set of policy tools available to promote science, technology and innovation. Innovation prizes, or prizes that involve some kind of technological innovation, are among those that policy makers are most interested in. To date however, despite the long history of prizes as incentives for science and technology, their notable potential, recent popularity and increasing policy interest, there has been little empirically based scientific knowledge on how to design, manage and evaluate innovation prizes.

This research has investigated technology prizes and the means by which they induce innovation or other effects related with technological development. The project was set out to engage four key aspects of prizes for which there have been significant knowledge gaps: the motivation of entrants, their R&D activities, their technology outputs and the overall effect of prizes on innovation. Using an empirical, multiple case-study methodology and multiple types of data sources, this research investigated three cases of recent aerospace technology prizes: a main case study, the Google Lunar X Prize (GLXP) for robotic Moon exploration; and two pilot cases, the Ansari X Prize (AXP) for the first private reusable manned spacecraft, and the Northrop Grumman Lunar Lander Challenge (NGLLC) for flights of reusable rocket-powered vehicles.

Prizes can be considered an incentive and support mechanism for technological development and also an indirect mechanism to influence industry and public perceptions about S&T issues. Prizes can selectively target certain technologies, R&D performers and geographic areas, induce increasing levels of R&D activity and under certain conditions induce technological innovation as well. Prizes can also leverage funding significantly when they have a widespread, decentralized participation but involve higher programmatic risks than other more traditional incentive tools such as R&D contracts and grants. Prizes can induce research, development, diffusion or commercialization of technologies and other related outcomes. Prizes however are not always the most appropriate incentive and their successful implementation requires many design parameters to be properly set.

The evidence gathered in this investigation shows that prizes can be considered a phenomenon with features that cannot be analyzed exclusively in economic terms. These empirical findings can inform more insightfully the process of implementation of prize competitions and particularly the process of calculation of the monetary reward and the design of other non-monetary incentives. This investigation also shows how different types of entrants respond to different types of incentives and suggests an opportunity to design prizes that target specific groups of individuals and organizations or potentially create problem-solving communities. Technology prizes offer varied incentives. Prize incentives are those created by the announcement and development of competitions and include the cash reward and other non-monetary benefits (e.g. reputation, visibility, opportunity to participate in technology development). Technology incentives are those linked to the market value of the technologies involved in competitions. All these incentives attract entrants with diverse characteristics. In modern technology prizes, non-monetary incentives in particular are more effective than other prize incentives to attract both unconventional entrants (i.e. individuals and organizations generally not involved with the prize technologies) and entrants that are familiar with the prize technologies. The latter however may use prizes as a step toward the creation of a commercial enterprise and make that their priority. The monetary reward is not as important as other prize incentives in the decision to enter prizes, yet it is still important to position the competition in the media and disseminate the idea of the prize more widely.

Prizes can induce increasing R&D activities to target various technological goals, but the evolution of prize competitions is generally uncertain. The overall organization of prize R&D activities depends on entrant-level factors such as entrants' goals, strategies and resources, and is not directly influenced by prize design. Moreover, some features of the organization of prize R&D activities are also observed in other instances of R&D in the same sector. The most remarkable characteristic of prize R&D activities is ultimately the iterative problem-solving efforts that entrants pursue to achieve the prize challenge. Prize design factors such as the time allowed to find solutions and the financial gap created by expensive challenges become part of the problem that entrants have to solve. Furthermore, the fundraising efforts of entrants may in some circumstances divert their R&D efforts. Prize technologies can range from experimental to marketable technologies but it is difficult for sponsors to focus that activity without hindering creativity. The evidence shows that unconventional entrants can contribute fresh ideas and new perspectives to solve problems but this ability is also shared with other entrants. Technological breakthroughs can only be enabled and not directly incentivized. Ultimately the characteristics of the prize technologies depend on prize designs, ongoing projects, current-day technologies and entrant-level factors.

Finally, in a context in which corporate and academic R&D choices are generally limited to other traditional incentive mechanisms, prizes can induce innovation over and above what would have occurred anyway, although their overall effect still depends significantly on the characteristics of the prize entrants and the evolution of the context of competitions. The ability of prizes to induce innovation is larger when there are larger prize incentives, more significant technology gaps implicit in the prize challenge and open-ended challenge definitions. Technology incentives also play a key role in prizes aimed at inducing technology implementation and commercialization. Intellectual property rights are particularly relevant in that case and complement the effect of prizes by enabling fundraising and commercialization. Under certain circumstances (such as unfavorable economic conditions and challenges of expensive achievement) other complementary incentives and support from prize sponsors (such as seed-funding up-front) may be needed to make prizes work.

There are at least two areas in which prizes are more likely to present advantages over other incentive mechanisms for innovation. Prizes can accelerate the development and/or commercialization of existing technologies that are held back for diverse reasons and help to leverage public money with external ideas, collaborative efforts and the participation of diverse individuals and organizations (including companies, universities, NGOs and others generally not involved with the prize technologies) and the public. To be effective however prizes have to target specific technological problems for which the achievement of a solution will be unequivocal, verifiable and visible to the judges, the competitors and the public; address issues that can be tackled with the base technologies that are generally available to all entrants and within a reasonable development lead time given by the prize deadline; and balance cash rewards with other non-monetary incentives that are also important for entrants.

The lack of previous research and empirical evidence on prize cases required making important decisions in terms of methodologies, data gathering and analysis for this investigation. Most importantly, there was the trade-off between the comprehensiveness of the study and the degree of detail and strength of insights provided by the evidence. The author decided to balance depth and breadth with the goal of seeking explanations of the entire innovation process induced by prizes and contributing empirically grounded insights for theory building. In terms of data gathering this investigation required creating new data gathering instruments and coordinating the collection of significant amounts of data from multiple sources. The analysis required an intense effort to be able to disentangle the effects induced by the prize from those related with the characteristics of entrants or the context of the competitions.

The author benefited from the opportunity of interviewing and observing entrants in an ongoing competition (the GLXP), which allowed the gathering of richer data but made it impossible to anticipate the ultimate outputs of the competition. The findings demonstrate that to better understand the impact of prizes on innovation, it is more important to learn first-hand the nature and extent of the effort of participants (and their partners and collaborators) and their perceptions than to describe the characteristics of the winning entry. The ideal assessment of a prize should comprise both approaches (direct observation and retrospective analysis) but this is sometimes impracticable when the investigation focuses on a long-term competition.

There are still many prize-specific implementation questions and considerations that remain to be addressed. For example: What is the most efficient balance between cash purses and other incentives offered by prizes in the context of different technological fields and industry sectors? What is the most appropriate balance between the degree of openness of the competition and the possibility of having only the most qualified entries? To what extent will the widespread use of prizes create prize-fatigue, and thus dilute the incentive power of individual competitions? Further empirical research should address this kind of question and substantiate our knowledge and prize theories. Future research should seek to develop better data sources and methods to quantify and qualify the magnitude of the collaborative effort that includes volunteers and partners. The approach followed by this investigation demonstrates that real-time data from ongoing competitions can be more insightful than historical accounts to study certain aspects of innovation in prizes. This is related with the notion that the most interesting effect of prizes is not only the winning entry but also the activities and outputs of runners-up, other entrants and other entities that participate only indirectly.

Prizes are an interesting instrument to incentivize innovation and a phenomenon that represents the importance of innovation and innovation policy in the highly competitive 21st century. Innovation has become a very valued attribute in the new economy and is generally considered by companies and governments a key factor to succeed in both traditional and high-technology fast-paced markets. Prize sponsors do more than offer rewards for a technological solution. Through this means they also position themselves as an authority in their field and as proactive contributors to the new economy. Prizes might play an important role in the creation of future science and technology policies that combine prizes and other more traditional incentive mechanisms to induce widely distributed efforts and seek technology-based solutions to pressing social and economic issues.

Appendix

Table A.1 AXP entrants

#	Team name	Created	Entered the competition in[a]	Number of members	Based in	Description[b]
1	Acceleration Engineering	N/A	1996	1	Bath, Michigan, USA	Volunteer team created to enter the competition
2	Advent Launch Services	N/A	1996	100	Houston, Texas, USA	Volunteer team that became an employee-owned corporation
3	Aeronautics and Cosmonautics Romanian Association (ARCA)	1999	2002	8	Ramnicu Valcea, Romania	Non-profit organization created by students
4	Armadillo Aerospace	2000	2002	6	Mesquite, Texas, USA	Newly created, independent R&D team
5	American Astronautics Corporation	N/A	2003	N/A	Oceanside, California, USA	Company created to enter the competition
6	Bristol Spaceplanes Ltd.	1991	1997	N/A	Bristol, England, UK	Consulting company that re-directed its activities
7	Canadian Arrow	N/A	2000	18	London, Ontario, Canada	Volunteer team with no experience in aerospace industry

Table A.1 (continued)

#	Team name	Created	Entered the competition in[a]	Number of members	Based in	Description[b]
8	Da Vinci Project	N/A	2000	14	Toronto, Ontario, Canada	Independent R&D team
9	Pablo de Leon & Associates	N/A	1997	6	Buenos Aires, Argentina	Company created to enter the competition
10	Discraft Corporation	N/A	1997	N/A	Portland, Oregon, USA	N/A
11	Flight Exploration	N/A	N/A	N/A	London, England, UK	N/A
12	Fundamental Technology Systems	N/A	2000	7	Orlando, Florida, USA	Company that re-directed its activities
13	HARC (High Altitude Research Corporation)	N/A	N/A	N/A	Huntsville, Alabama	N/A
14	IL Aerospace Technologies	N/A	2002	7	Zichron Ya'akov, Israel	Company created to enter the competition
15	Interorbital Systems	1996	2003	8	Mojave, California, USA	Pre-existing aerospace company
16	Kelly Space and Technology	N/A	N/A	N/A	San Bernadino, California, USA	Pre-existing aerospace company
17	Lone Star Space Access Corporation	N/A	N/A	N/A	Houston, Texas, USA	N/A

18	Micro-Space Inc.	N/A	2003	6	Denver, Colorado, USA	Company that re-directed its activities
19	PanAero Inc.	1997	1997	9	Fairfax, VA, USA	Newly created aerospace engineering company
20	Pioneer Rocketplane Inc. (now Rocketplane Kistler)	2001	N/A	N/A	Oklahoma City, OK	Company created to enter the competition
21	Scaled Composites	N/A	2001	135 employees	Mojave, California, USA	Pre-existing aircraft design firm
22	Space Transport Corporation	2002	2003	N/A	N/A	Company created to enter the competition
23	Starchaser Industries	1998	1996	35	Cheshire, England, UK	Research foundation later incorporated as a company
24	Suborbital Corporation	N/A	N/A	N/A	Moscow, Russia	N/A
25	TGV Rockets	N/A	1999	6	Bethesda, Maryland, USA	Company created to enter the competition
26	Vanguard Spacecraft	N/A	2003	6	Bridgewater, Massachusetts, USA	Company created to enter the competition

Notes:

a. Number of members is as of 2003 (otherwise indicated)

b. Teams were classified into conventional and unconventional from the point of view of the similarity of their pre-prize activities as a group or organization in relation to the prize challenge.

Source: Different sources cited in the text and references and analysis of the author.

Table A.2 NGLLC entrants (2006–09)

#	Team name	Created	NGLLC participation	Number of members[a]	Based in	Description
1	Masten Space Systems	2004	2007, 2009 (2nd place Level I, 1st place Level II)	5	Mojave, CA	Small startup, rocketry and propulsion company
2	Acuity Technologies	1992	2006, 2007, 2008	5	Menlo Park, CA	Pre-existing, robotics company that re-directed its activities
3	Micro-Space	Before 2004	2006, 2007	3	Denver, CO	Pre-existing company in high-tech design that re-directed its activities
4	Armadillo Aerospace	2000	2006, 2007, 2008 (1st place Level I), 2009 (2nd place Level II)	8	Mesquite, TX	Independent R&D team
5	BonNova	N/A	2007, 2008, 2009 (withdrawn)	6	Tarzana, CA	Small design company with aerospace and other industries experience
6	High Expectations Rocketry	N/A	2008	4	Moscow, ID	Small engineering research group that re-directed its activities

214

#	Name	Founded	Years	Members	Location	Description
7	Paragon Labs	2000	2007, 2008	9	Denver, CO	Pre-existing aerospace consultancy firm (re-directed activities to develop suborbital launch vehicles and VTOL technologies)
8	Speed Up	N/A	2007	1	N/A	Independent group with support from a private company, Frontier Astronautics
9	Phoenicia	N/A	2008	5	Emeryville, CA	Independent group created to compete for the prize
10	Seraphim Works	N/A	2008	N/A	N/A	N/A
11	TrueZer0	N/A	2008	4	Chicago, IL	Pre-existing engineering company that re-directed its activities
12	Unreasonable Rocket	N/A	2007, 2008, 2009	2	Solana Beach, CA	Father and son amateur team

Note: a. Number of members is as of 2008 (or 2007 if data for 2008 were not found).

Source: Different sources cited in the text and references and analysis of the author.

Table A.3   GLXP entrants

Team name	Type of entity	Reported country in GLXP site	Created for GLXP	Entered GLXP	Withdrawn
Odyssey Moon	For-profit	Multi-natl./Isle of Man	No	Dec-07	–
ARCA	Non-profit	Romania	No	Feb-08	–
Chandah	Independent	USA	Yes	Feb-08	Jan-11
Astrobotic	For-profit	USA	Yes	Feb-08	–
LUNARecon (Lunatrex)	For-profit	USA	No	Feb-08	Dec-09
SCSG	Independent	USA	Yes	Feb-08	Jun-08
Micro-space	For-profit	USA	No	Feb-08	Nov-10
Italia	N/A	Italy	Yes	Feb-08	–
Frednet	Non-profit	Multi-national	Yes	Feb-08	–
Quantum3	For-profit	USA	Yes	Feb-08	Feb-09
Selene	Independent	China, Germany	Yes	May-08	–
Stellar	For-profit	USA	Yes	May-08	–
Jurban	Part of larger org. (Non-profit)	USA	Yes	May-08	–
Advaeros	For-profit	Multi-national	No	May-08	Nov-10
Independence X	Part of larger org. (Non-profit)	Malaysia	Yes	Sep-08	–
Omega Envoy	Non-profit	USA	No	Oct-08	–
Next Giant Leap	For-profit	USA	Yes	Dec-08	–
Euroluna	Non-profit	Danish, Swiss, Italian	Yes	Dec-08	–
Synergy Moon	N/A	Multi-national	Yes	Feb-09	–

Team	Type	Country		Date	
White Label Space	Non-profit	Multi-national	Yes	May-09	–
Part Time Scientists	For-profit	Germany	Yes	Jun-09	–
Selenokhod	For-profit	Russia	No	Sep-09	–
C-Base Open Moon	Non-profit	Germany	Yes	Oct-09	Jul-11
Barcelona Moon	For-profit	Spain	Yes	Apr-10	–
Rocket City Space Pioneers	Part of larger org. (For-profit)	USA	Yes	Sep-10	–
Moon Express	For-profit	USA	No	Oct-10	–
Teams that entered after data gathering ended					
Team Space IL	Non-profit	Israel	Yes	Jan-11	–
Mystical Moon	N/A	Multi-national	Yes	Feb-11	Jul-11
Team Puli	Independent	Hungary	Yes	Feb-11	–
SpaceMETA team	Independent	Brazil	Yes	Feb-11	–
Plan B	For-profit	Canada	No	Feb-11	–
Penn State Lunar Lion Team	Non-profit	USA	Yes	Feb-11	–
Angelicum	For-profit	Chile	No	Feb-11	–
Team Indus	For-profit	India	Yes	Feb-11	–
Team Phoenicia	For-profit	USA	No	Feb-11	–

Source: GLXP official website and questionnaire to GLXP teams.

References

Abramowicz, M. (2003). Perfecting patent prizes. *Vanderbilt Law Review*, **56**(1), 114–236.

Academy of Achievement. (1991). Paul MacCready Interview. Retrieved May 1, 2009, from http://www.achievement.org/autodoc/page/mac0int-2

Aerospace Safety Advisory Panel (ASAP). (2011). ASAP Annual Report for 2010. Washington, DC: Aerospace Safety Advisory Panel.

Amabile, T.M., DeJong, W., and Lepper, M.R. (1976). Effects of externally imposed deadlines on subsequent intrinsic motivation. *Journal of Personality and Social Psychology*, **34**(1), 92–8.

Anastas, P.T., and Zimmerman, J.B. (2007). *Why We Need a Green Nano Award & How to Make it Happen*. Woodrow Wilson International Center for Scholars.

Anton, J.J., and Yao, D.A. (1990). Measuring the effectiveness of competition in defense procurement: a survey of the empirical. *Journal of Policy Analysis & Management*, **9**(1), 60–79.

Armadillo Aerospace. (2008). Engine Work, Methane Work, Selling Vehicles, Lynx. Retrieved June 22, 2010, from http://armadilloaerospace.com/n.x/Armadillo/Home/News?news_id=357

Astrobotic Technology. (2010). Tranquility Trek. Retrieved December 12, 2010, from http://astrobotic.net/activities/tranquility-trek/

Bacharach, S.B. (1989). Organizational theories: some criteria for evaluation. *Academy of Management Review*, **14**(4), 496–515.

Bain, P.G., Mann, L., and Pirola-Merlo, A. (2001). The innovation imperative: the relationships between team climate, innovation, and performance in research and development teams. *Small Group Research*, **32**(1), 55–73.

Baird, F., Moore, C.J., and Jagodzinski, A.P. (2000). An ethnographic study of engineering design teams at Rolls-Royce Aerospace. *Design Studies*, **21**(4), 333–55.

Balint, T.S., Kolawa, E.A., Cutts, J.A., and Peterson, C.E. (2008). Extreme environment technologies for NASA's robotic planetary exploration. *Acta Astronautica*, **63**(1–4), 285–98.

Belfiore, M. (2007). *Rocketeers. How a Visionary Band of Business*

Leaders, Engineers, and Pilots is Boldly Privatizing Space. New York, NY: HarperCollins Publishers.

Best, J. (2008). Prize proliferation. *Sociological Forum,* **23**, 1–27.

Bitten, R.E. (2008). Perspectives on NASA mission cost and schedule performance trends. Retrieved February 2, 2011, from http://spirit. as.utexas.edu/~fiso/telecon/Bitten_7-02-08.pdf

Bloomberg Businessweek. (2011). Astrobotic Technology, Inc.: Private Company Information. Retrieved February 23, 2011, from http://investing.businessweek.com/research/stocks/private/snapshot.asp?privcapId=39444578

Boeing, Martin Marietta, General Dynamics, MacDonnell Douglas, Lockheed, & Rockwell. (1994). *Commercial Space Transport Study Final Report (CSTS).* Sponsored by NASA's Langley Research Center.

Bond, S., Harnoff, D., and Van Reenen, J. (1999). Investment, R&D and financial constraints in Britain and Germany. Working Paper Series, Institute For Fiscal Studies, **99**(5).

Bonin, G. (2009). Microspace and human spaceflight. Retrieved February 2011, from http://www.thespacereview.com/article/1441/1

Bower, J.L., and Christensen, C.M. (1995). Disruptive technologies: catching the wave. *Harvard Business Review,* **73**, 43–53.

Boyle, A. (2004). 'Spaceship team gets its $10 million prize.' Retrieved Feb 19, 2008, from http://www.msnbc.msn.com/id/6421889/

Brabham, D.C. (2008). Crowdsourcing as a model for problem solving. *Convergence: The International Journal of Research into New Media Technologies,* **14**(1), 75–90.

Braun, R.D. (2010). NASA Innovation and Technology Preliminary Planning. Retrieved February 2, 2011, from http://www.spacepolicyonline.com/pages/images/stories/Braun_-_NASA_OCT_March_9_ASEB.pdf

Broda-Bahm, K.T. (1996). Counterfactual problems: addressing difficulties in the advocacy of counter-to-fact causal claims. *Contemporary Argumentation and Debate,* **17**, 19–31.

Bromberg, J.L. (2000). *NASA and the Space Industry.* Baltimore and London: The Johns Hopkins University Press.

Brown, P.J. (2010). India, Russia squeeze Google Moon racers. Retrieved February 2, 2011, from http://www.atimes.com/atimes/China/LH12Ad03.html

Brunt, L., Lerner, J., and Nicholas, T. (2008). *Inducement Prizes and Innovation.* NBER.

Bugos, G.E., and Boyd, J.W. (2008). Accelerating entrepreneurial space: the case for an NACA-style organization. *Space Policy,* **24**(3), 140–47.

Bullinger, A.C., Neyer, A.-K., Rass, M., and Moeslein, K.M. (2010).

Community-based innovation contests: where competition meets cooperation. *Creativity and Innovation Management*, **19**(3), 290–303.

Byko, M. (2004). SpaceShipOne, the Ansari X Prize, and the materials of the civilian space race. *JOM*, **56**(11), 24–28.

California Space Education and Workforce Institute. (2009). Regolith Excavation Challenge. Retrieved February 2, 2011, from http://regolith. csewi.org/

Charlton, B.G. (2007). Mega-prizes in medicine: big cash awards may stimulate useful and rapid therapeutic innovation. *Medical Hypotheses*, **68**(1), 1–3.

Che, Y.-K., and Gale, I. (2003). Optimal design of research contests. *American Economic Review*, **93**(3), 646–71.

Chesbrough, H.W. (2003). *Open Innovation: The New Imperative for Creating and Profiting from Technology*. Harvard Business Press.

Chesbrough, H.W. (2006). *Open Business Models. How to Thrive in the New Innovation Landscape*. Boston, MA: Harvard Business School Press.

Cisco. (2010). Cisco Announces Winner of Global I-Prize Innovation Competition. Retrieved September 19, 2010, from http://newsroom. cisco.com/dlls/2010/prod_062910b.html

Clark, R.E. (2003). Fostering the work motivation of individuals and teams. *Performance Improvement*, **42**(3), 21–9.

Cohen, L.R., and Noll, R.G. (1991). *The Technology Pork Barrel*. Washington, DC: The Brookings Institution.

Cohen, W.M., Nelson, R.R., and Walsh, J.P. (2000). Protecting their intellectual assets: appropriability conditions and why U.S. manufacturing firms patent (or not). NBER Working Paper No. w7552.

Collins, P.Q., and Ashford, D.M. (1986). Potential economic implications of the development of space tourism. Presented at 37th IAF Congress. Innsbruck.

Congressional Budget Office (CBO). (2004). A Budgetary Analysis of NASA's New Vision for Space Exploration.

Congressional Budget Office (CBO). (2007). Federal Support for Research and Development.

Copenhagen Suborbitals. (2011). Copenhagen Suborbitals. Retrieved February 2, 2011, from http://www.copenhagensuborbitals.com/

Courtland, R. (2009). Contenders square up in battle of the lunar landers. Retrieved April 24, 2010, from http://www.newscientist.com /article/dn18043-contenders-square-up-in-battle-of-the-lunar-landers. html

Creswell, J.W., and Plano Clark, V.L. (2007). *Designing and Conducting Mixed Methods Research*. Thousand Oaks: SAGE Publications.

Cronin, M. (2011). Fly me to the moon. *The Daily*. Retrieved February 6, 2011, from http://www.thedaily.com/page/2011/02/06/020611-news-lunar-mining-1of5/

Crosland, M., and Galvez, A. (1989). The emergence of research grants within the prize system of the French Academy of Sciences, 1795–1914. *Social Studies of Science*, **19**(1), 71–100.

Cucit, L., Nosella, A., Petroni, G., and Verbano, C. (2004). Management and organizational models of the European Space Agencies: the results of an empirical study. *Technovation*, **24**(1), 1–15.

Culver, L., Escudero, L., Grindle, A., Hamilton, M., and Sowell, J. (2007). Policies, incentives, and growth in the newspace industry. Working Paper.

Daniel, W.W. (1975). Nonresponse in sociological surveys. A review of some methods for handling the problem. *Sociological Methods & Research*, **3**(3), 291–307.

Danto, E.A. (2008). *Historical Research*. New York: Oxford University Press.

DARPA. (2006). Report to Congress. DARPA Prize Authority. Fiscal Year 2005 report in accordance with 10 U.S.C. § 2374a: DARPA.

DARPA. (2008). DARPA Urban Challenge. Fiscal Year 2007 Report.

Dasgupta, P., and Stiglitz, J. (1980). Uncertainty, industrial structure, and the speed of R&D. *The Bell Journal of Economics*, **11**(1), 1–28.

Davidian, K. (2005). Prize Competitions and NASA's Centennial Challenges Program, International Lunar Conference 2005: DMG Associates.

Davidian, K. (2007). *Prizes, Prize Culture, and NASA's Centennial Challenges*. DMG Associates under contract to the National Aeronautics and Space Administration, Headquarters, Washington DC, USA 20546-0001.

Davidian, K. (2010). Interview with Ken Davidian, Director of Research at the FAA Office of Commercial Space Transportation (AST), September 20, 2010.

Davis, L., and Davis, J. (2004). How effective are prizes as incentives to innovation? Evidence from three 20th century contests. DRUID Summer Conference 2004. Elsinore, Denmark.

Davis, L.N. (2002). Should we consider alternative incentives for basic research? Patents vs. prizes. *Industrial Dynamics of the New and Old Economy – Who Is Embracing Whom?* Copenhagen/Elsinore.

de Laat, E.A.A. (1997). Patents or prizes: monopolistic R&D and asymmetric information. *International Journal of Industrial Organization*, **15**(3), 369–390.

Diamandis, P. (2004). NASA contests and prizes: How can they help advance space exploration? US Congress, 108th Congress, House Committee on Science, 2nd Session. Retrieved February 4, 2010, from http://commdocs.house.gov/committees/science/hsy94832.000/hsy94832_0.HTM

Diamandis, P. (2008). Google Lunar X PRIZE – The BlastOff Story. Retrieved February 3, 2011, from http://www.youtube.com/watch?v=KfA6hLj2j5U.

Diamandis, P. (2009). Using incentive prizes to drive creativity, innovation and breakthroughs. Retrieved in Fall 2009 from http://ocw.mit.edu/courses/engineering-systems-division/esd-172j-x-prize-workshop-grand-challenges-in-energy-fall-2009/readings/MITESD_172JF09_Diamandis.pdf.

Dillman, D.A. (2000). *Mail and Internet Surveys. The Tailored Design Method.* (2nd edn). New York: John Wiley & Sons, Inc.

Discovery Channel. (2005). *Black Sky: The Race for Space & Winning the X-Prize.* 2 DVD Set.

Dornheim, M.A. (2003). Can $$$ buy time? *Aviation Week & Space Technology*, **158**(21), 56.

Dosi, G., and Egidi, M. (1991). Substantive and procedural uncertainty. *Journal of Evolutionary Economics.* Springer Science & Business Media B.V.

Eisenhardt, K.M. (1989). Building theories from case study research. *Academy of Management Review.* Academy of Management.

English, J.F. (2005). *Prizes, Awards, and the Circulation of Cultural Value.* Cambridge, MA: Harvard University Press.

Evadot. (2009). Evadot Podcast #4 – An Interview with Team FredNet. Retrieved February 2, 2011, from http://evadot.com/2009/07/01/evadot-podcast-4-%E2%80%93-an-interview-with-team-frednet/

Fink, W., Dohm, J.M., Tarbell, M.A., Hare, T.M., and Baker, V.R. (2005). Next-generation robotic planetary reconnaissance missions: a paradigm shift. *Planetary and Space Science*, **53**(14–15), 1419–26.

Fullerton, R.L., and McAfee, R.P. (1999). Auctioning entry into tournaments. *Journal of Political Economy*, **107**(3), 573–605.

Futron Corporation. (2010a). *Commercial Lunar Transportation Study.* Market Assessment Summary.

Futron Corporation. (2010b). Emerging commercial lunar activities: assessing market size and development. Presentation to the Google Lunar X Prize Summit. Isle of Man, UK.

Gallini, N., and Scotchmer, S. (2001). *Intellectual Property: When Is It the Best Incentive System?* Berkeley: University of California.

Gans, J.S., and Stern, S. (2010). Is there a market for ideas? *Industrial and Corporate Change*, **19**(3), 805–37.

Garcia, R., and Calantone, R. (2002). A critical look at technological innovation typology and innovativeness terminology: a literature review. *Journal of Product Innovation Management*, **19**(2), 110–32.

Goldsmith, R. (2009). Team LunaTrex Talk to the Space Fellowship About Rockets, Rovers and GLXP Rules. Retrieved February 1, 2011, from http://spacefellowship.com/news/art11617/team-lunatrex-talk-to-the-space-fellowship-about-rockets-rovers-and-glxp-rules.html

Google Lunar X Prize website (GLXP). (2009). GLXP Forum. Retrieved February 2, 2011, from http://www.googlelunarxprize.org/lunar/forum-glxp

Google Lunar X Prize website (GLXP). (2010a). Isle of Man Summit: Rocket City Space Pioneers Presentation video. Retrieved January 15, 2011, from http://www.googlelunarxprize.org/lunar/teams/rocket-city-space-pioneers/blog/isle-of-man-summit-rocket-city-space-pioneers-presentation

Google Lunar X Prize website (GLXP). (2010b). Synergy Moon. Retrieved February 2, 2011, from http://www.googlelunarxprize.org/lunar/teams/synergy-moon

Google Lunar X Prize website (GLXP). (2011a). RoverX Development Has Been Started. Retrieved February 2, 2011, from http://www.googlelunarxprize.org/lunar/teams/selene/blog/roverx-development-has-been-started

Google Lunar X Prize website (GLXP). (2011b). Update on TALARIS. Retrieved March 8, 2011, from http://www.googlelunarxprize.org/lunar/teams/next-giant-leap/blog/update-on-talaris

Greason, J. (2010). Interview with Jeff Greason, XCOR, November 9, 2010.

Green, J.R., and Scotchmer, S. (1995). On the division of profit in sequential innovation. *RAND Journal of Economics*, **26**, 20–33.

Griffin, M.D. (2007). Letter to Dr. Peter H. Diamandis, Chairman and Chief Executive Officer, X PRIZE Foundation. Retrieved February 2, 2011, from http://www.googlelunarxprize.org/files/downloads/lunar/nasa_letter.pdf

Grishagin, V.A., Sergeyev, Y.D., and Silipo, D.B. (2001). Firms' R&D decisions under incomplete information. *European Journal of Operational Research*, **129**(2), 414–33.

Gump, D.P. (1990). *Space Enterprise: Beyond NASA*. New York, NY: Praeger Publishers.

Gupta, A.K., and Wilemon, D. (1990). Accelerating the development of technology-based new products. *California Management Review*, Winter, 24–44.

Hall, B.H. (1992). Investment and research and development at the firm level: does the source of financing matter? NBER Working Paper Series, 4096.

Hall, B.H. (2002). The financing of research and development. *Oxford Review of Economic Policy*, **18**(1), 35–51.

Hawkins, D.F. (1975). Estimation of nonresponse bias. *Sociological Methods & Research*, **3**(4), 461–88.

Himmelberg, C.P., and Petersen, B.C. (1994). R&D and internal finance: a panel study of small firms in high-tech industries. *Review of Economics & Statistics*, MIT Press.

Homans, C. (2010). The wealth of constellations. *Washington Monthly* (May/June 2010), 18–26.

Horrobin, D.F. (1986). Glittering prizes for research support. *Nature*, **324**, 221.

Hsu, J. (2010). NASA plans new robot generation to explore moon, asteroids. Retrieved January 21, 2011, from http://www.space.com/8157-nasa-plans-robot-generation-explore-moon-asteroids.html.

Hudgins, E.L. (ed.). (2002). *Space: The Free-market Frontier*. Washington, DC: Cato Institute.

Hudson, F. (2008). Gravity is not the main obstacle for America's space business. Government is. Retrieved February 2, 2011, from http://www.economist.com/node/11965352?story_id=11965352.

Hulsheger, U.R., Anderson, N., and Salgado, J.F. (2009). Team-level predictors of innovation at work: a comprehensive meta-analysis spanning three decades of research. *Journal of Applied Psychology*, **94**(5), 1128–45.

Hutter, K., Hautz, J., Füller, J., Mueller, J., and Matzler, K. (2011). Communitition: the tension between competition and collaboration in community-based design contests. *Creativity and Innovation Management*, **20**(1), 3–21.

iMARS. (2008). Preliminary Planning for an International Mars Sample Return Mission. Retrieved February 2, 2011, from http://mepag.jpl.nasa.gov/reports/iMARS_FinalReport.pdf

Indian Space Research Organization (ISRO). (2010). What is Chandrayaan-2. Retrieved February 2, 2011, from http://www.chandrayaan-i.com/index.php/chandrayaan-2.html

Kalil, T. (2006). *Prizes for Technological Innovation*. Washington, DC: The Brookings Institution.

Karau, S.J., and Kelly, J.R. (1992). The effects of time scarcity and time abundance on group-performance quality and interaction process. *Journal of Experimental Social Psychology*, **28**(6), 542–71.

Kay, L. (2010). Notes from team visits and 4th annual GLXP Summit (Isle of Man, UK).

Kay, L. (2011). Managing innovation prizes in government. IBM Center for the Business of Government.

Kemp, K. (2007). *Destination Space. How Space Tourism Is Making Science Fiction A Reality*. London: Virgin Books.

Kessler, E.H., and Chakrabarti, A.K. (1999). Speeding up the pace of new product development. *Journal of Product Innovation Management*, 16(3), 231–47.

Kessner, T. (2010). *The Flight of the Century: Charles Lindbergh and the Rise Of American Aviation*. Oxford: Oxford University Press.

Kieff, S.F. (2001). Property rights and property rules for commercializing inventions. *Minnesota Law Review*, 85, 697–754.

Kleemann, F., Voß, G.G., and Rieder, K. (2008). Un(der)paid innovators: the commercial utilization of consumer work through crowdsourcing. *Science, Technology & Innovation Studies*, 4(1), July.

Kleiman, M.J. (2010). Licensing intellectual property rights out of this world. Retrieved January 21, 2011, from http://www.spacenews.com/commentaries/100901-blog-licensing-intellectual-property-rights.html

Knowledge Ecology International (KEI). (2008). Selected Innovation Prizes and Reward Programs. Retrieved February 5, 2011, from http://keionline.org/misc-docs/research_notes/kei_rn_2008_1.pdf

Kolodny, L. (2011). Move over, Rover: next giant leap gets $1 million grant to build hopping moon landers. Retrieved January 21, 2011, from http://techcrunch.com/2011/01/21/next-giant- leap-gets-1-million-grant-to-build-hopping-moon-landers/

Kranz, G. (2000). *Failure Is Not An Option*. New York: Simon & Schuster.

Kremer, M. (1998). Patent buyouts: a mechanism for encouraging innovation. *Quarterly Journal of Economics*, 113(4), 1137–67.

Kremer, M. (2000). Creating markets for new vaccines. Part I: Rationale. NBER Working Paper: National Bureau of Economic Research.

Levin, R.C., Klevorick, A.K., Nelson, R.R., Winter, S.G., Gilbert, R., and Griliches, Z. (1987). Appropriating the returns from industrial research and development. *Brookings Papers on Economic Activity*, 1987(3), 783–831.

Linehan, D. (2008). *SpaceShipOne. An Illustrated History*. Minneapolis, MN: Zenith Press.

Locke, E.A., and Latham, G.P. (1990). *A Theory of Goal Setting and Task Performance*. New Jersey: Prentice-Hall.

LunaCorp. (1996). Payfor – LunaCorp. Retrieved February 1, 2011, from http://web.archive.org/web/19961109200440/www.lunacorp.com/payfor.html

Macauley, M.K. (2005). Advantages and disadvantages of prizes in a portfolio of financial incentives for space activities. *Space Policy*, 21(2), 121–8.

MacDonald, A., and Marshall, W.S. (2008). The common spacecraft bus and lunar commercialization. Retrieved February 2, 2011, from http://commercialspace.pbworks.com/f/Public+HTV.pdf

MacLeod, R.M. (1971). Of medals and men: a reward system in Victorian science, 1826–1914. *Notes and Records of the Royal Society of London*, **26**(1), 81–105.

Mankins, J.C. (1995). Technology readiness levels. NASA Office of Space Access and Technology.

Mansfield, E. (1988). The speed and cost of industrial innovation in Japan and the United States – external vs internal technology. *Management Science*, **34**(10), 1157–68.

Marsh, G. (2011). Interview with George Marsh, Retired Lockheed Martin Space Systems Co. Executive Vice President, January 19, 2011.

Marshall, W.S., Turner, M.F., Butler, P.H., and Weston, A.R. (2007). Small spacecraft in support of the lunar exploration program. A study effort of the Small Spacecraft Office, NASA-Ames Research Center, Moffett Field, California, 94035, USA.

Maryniak, G. (2005). When will we see a golden age of spaceflight? *Space Policy*, **21**, 111–19.

Maryniak, G. (2010). Interview with Gregg Maryniak, X Prize Foundation, October 27, 2010.

Masten Space Systems. (2009). Masten Space Systems Qualifies for $1 Million Prize. Retrieved March 4, 2010, from http://masten-space.com/blog/?p=485

Masten Space Systems. (2010). Personnel Updates. Retrieved February, 2011, from http://masten-space.com/2010/09/10/personnel-updates/

Masters, W.A. (2003). Research prizes: a mechanism to reward agricultural innovation in low-income regions. *AgBioForum*, **6**(1&2), 71–4.

Masters, W.A., and Delbecq, B. (2008). Accelerating innovation with prize rewards: history and typology of technology prizes and a new contest design for innovation in African agriculture. *IPRI Conference on Advancing Agriculture in Developing Countries*. Addis Ababa.

Maurer, S.M., and Scotchmer, S. (2004). Procuring knowledge. In G. Libecap (ed.), *Intellectual Property and Entrepreneurship: Advances in the Study of Entrepreneurship, Innovation and Growth* (Vol. 15). The Netherlands: JAI Press (Elsevier), pp. 1–31.

Maxwell, J.A., and Miller, B.A. (2008). Categorizing and connecting strategies in qualitative data analysis. In S. Hesse-Biber and P. Leavy (eds), *The Handbook of Emergent Methods*. New York and London: The Guildford Press.

McCurdy, H.E. (1994). *Inside NASA. High Technology and Organizational Change in the U.S. Space Program*. Baltimore and London: The Johns Hopkins University Press.

McDowell, W.H. (2002). *Historical Research. A Guide.* London: Pearson Education Limited.

McKinsey & Company. (2009). *"And the winner is. . ." Capturing the promise of philanthropic prizes.*

Merton, R.K. (1973). *The Sociology of Science. Theoretical and Empirical Investigations.* Chicago: The University of Chicago Press.

Miles, M.B., and Huberman, A.M. (1994). *Qualitative Data Analysis: An Expanded Sourcebook.* Thousand Oaks, CA: SAGE Publications.

Mokyr, J. (2009). Intellectual property rights, the Industrial Revolution, and the beginnings of modern economic growth. *American Economic Review: Papers & Proceedings*, **99**(2), 349–55.

Moldovanu, B., and Sela, A. (2001). The optimal allocation of prizes in contests. *American Economic Review*, **91**(3), 542–58.

Morring, F. (2009). Masten building on X-Prize. Retrieved February 23, 2010, from http://www.aviationweek.com/aw/generic/story.jsp?id=news/xprize110909.xml&headline=Masten%20Building%20On%20X-Prize%20&channel=space

Mowery, D.C., Nelson, R.R., and Martin, B.R. (2010). Technology policy and global warming: why new policy models are needed (or why putting new wine in old bottles won't work). *Research Policy*, **39**(8), 1011–23.

MSNBC.com. (2007). NASA Extends Mars Rovers' Mission. Retrieved February 2011, from http://www.msnbc.msn.com/id/21327647/

Nalebuff, B.J., and Stiglitz, J.E. (1983). Prizes and incentives: towards a general theory of compensation and competition. *The Bell Journal of Economics*, **14**(1), 21–43.

NASA. (1997). Rangers and Surveyors to the Moon. Pasadena, CA: NASA Jet Propulsion Laboratory.

NASA. (2009a). Lunar Lander Challenge. Retrieved January 27, 2010, from http://www.nasa.gov/offices/ipp/innovation_incubator/centennial_challenges/lunar_lander/index.html

NASA. (2009b). Mars Exploration Rover Mission: Overview. Retrieved February 2, 2011, from http://marsrovers.jpl.nasa.gov/overview/

NASA. (2009c). Masten and Armadillo Claim Lunar Lander Prizes. Retrieved March 10, 2010, from http://www.nasa.gov/offices/ipp/innovation_incubator/centennial_challenges/cc_ll_feature_lvl2.html

NASA. (2010a). CCDev Information. Retrieved February 2, 2011, from http://www.nasa.gov/offices/c3po/partners/ccdev_info.html

NASA. (2010b). Fiscal Year 2011 Budget Estimates. Retrieved February 2, 2011, from http://www.nasa.gov/pdf/420990main_FY_201_%20Budget_Overview_1_Feb_2010.pdf

NASA. (2010c). Mars Sample Return Discussions. Retrieved February

228 *Technological innovation and prize incentives*

2, 2011, from http://www.spacepolicyonline.com/pages/images/stories/
PSDS%20Steering%20Cmte%20Feb%202010%20Li-Hayati.pdf
NASA. (2010d). NASA Awards Contracts For Innovative Lunar
Demonstrations Data. Retrieved February 2, 2011, from http://www.
nasa.gov/home/hqnews/2010/oct/HQ_10-259_ILDD_Award.html
NASA. (2010e). NASA Selects Two Firms for Experimental Space
Vehicle Test Flights. Retrieved August 30, 2010, from http://www.nasa.
gov/home/hqnews/2010/aug/HQ_10-203_CRuSR_Awards.html
NASA. (2010f). Solar System Exploration: Missions: By Target: Mars:
Past: Mars Pathfinder/Sojourner. Retrieved February 2, 2011, from
http://solarsystem.nasa.gov/missions/profile.cfm?MCode=Pathfinder&
Display=ReadMore
NASA. (2010g). Surveyor 7. Retrieved February 2, 2011, from http://
nssdc.gsfc.nasa.gov/nmc/spacecraftDisplay.do?id=1968-001A
NASA. (2011). NASA – Space Technology Roadmaps (DRAFT).
Retrieved February 2, 2011, from http://www.nasa.gov/offices/oct/
home/roadmaps/index.html
NASA. (2012). NASA SBIR & STTR: First Time Participant. Retrieved
January 13, 2012, from http://sbir.gsfc.nasa.gov/SBIR/ftp.html
National Academy of Engineering (NAE). (1999). *Concerning Federally
Sponsored Inducement Prizes in Engineering and Science*. Washington,
DC: National Academy of Engineering.
National Research Council (NRC). (2007). *Innovation Inducement Prizes
At The National Science Foundation* (No. 0-309-10465-3). Washington,
DC: The National Academies Press.
National Research Council (NRC). (2008). *A Constrained Space
Exploration Technology Program: A Review of NASA's Exploration
Technology Development Program*. Washington, DC.
Newell, R.G., and Wilson, N.E. (2005). Technology prizes for climate
change mitigation, Discussion Paper. Washington, DC: Resources For
The Future.
Northrop Grumman. (2007). Northrop Grumman Helps NASA Shape
Plans for Affordable Lunar Lander. Retrieved March 3, 2010, from http://
www.irconnect.com/noc/press/pages/news_releases.html?d=122412
O'Sullivan, A. (2003). Dispersed collaboration in a multi-firm, multi-team
product-development project. *Journal of Engineering and Technology
Management*, **20**(1–2), 93–116.
O'Sullivan, M.A. (2009). Funding new industries: a historical perspective
on the financing role of the U.S. stock market in the twentieth century.
In N.R. Lamoreaux and K.L. Sokoloff (eds), *Financing Innovation
in the United States, 1870 to the Present*. Cambridge, MA: The MIT
Press.

Odyssey Moon. (2008). Preparing for Moon 2.0. Retrieved October 29, 2010, from http://www.lpi.usra.edu/meetings/leagilewg2008/presenta tions/YLE/richards.pdf

OECD/Eurostat. (1997). *Proposed Guidelines for Collecting and Interpreting Technological Innovation Data – Oslo Manual*. Paris: OECD.

Parabolic Arc. (2010). NASA SBIR Program Funds Mars Sample Return Technologies. Retrieved February 1, 2011, from http://www. parabolicarc.com/2010/12/30/nasa-sbir-program-funds-mars-sample-return-technologies/

Part Time Scientists (PTS). (2011a). Part-Time-Scientists. Retrieved March 3, 2011, from http://www.part-time-scientists.com/

Part Time Scientists (PTS). (2011b). Third Fan Friday. Retrieved January 28, 2011, from http://www.part-time-scientists.com/2011/01/28/third-fan-friday/

Pedersen, L., Kortenkamp, D., Wettergreen, D., Nourbakhsh, I., and Smith, T. (2002). *NASA Exploration Team (NEXT) Space Robotics Technology Assessment Report*. Moffett Field, CA: NASA.

Penin, J. (2005). Patents versus ex post rewards: a new look. *Research Policy*, **34**, 641–56.

Petroni, G., Venturini, K., Verbano, C., and Cantarello, S. (2009). Discovering the basic strategic orientation of big space agencies. *Space Policy*, **25**(1), 45–62.

Polanyi, M. (1944). Patent reform. *The Review of Economic Studies*, **11**(2), 61–76.

Pomerantz, W. (2006). Advancements through prizes. NIAC Annual Meeting.

Pomerantz, W. (2007). NGLLC: early returns. Retrieved February 1, 2010, from http://www.xprize.org/blogs/wpomerantz/ngllc-early-returns

Pomerantz, W. (2010a). Interview with William Pomerantz, Senior Director of Space Prizes, X Prize Foundation, September 17, 2010.

Pomerantz, W. (2010b). Lessons from NASA's centennial challenges. Retrieved May 28, 2010, from http://thelaunchpad.xprize.org/2010/03/lessons-from-nasas-centennial.html

Pomerantz, W. (2011a). E-mail communication.

Pomerantz, W. (2011b). Will Pomerantz: a final Q&A. Retrieved February 22, 2011, from http://thelaunchpad.xprize.org/2011/02/will-pomerantz-final-q.html

Poniatowski, K.S., and Osmolovsky, M.G. (1995). Capabilities, costs, and constraints of space transportation for planetary missions. *Acta Astronautica*, **35**(Supplement 1), 587–96.

Reichhardt, T. (2008). Finding Apollo. Retrieved February 2, 2011, from http://www.airspacemag.com/space-exploration/Finding_Apollo.html

Ridenoure, R., and Polk, K. (1999). Private, commercial and student-oriented low-cost deep-space missions: a global survey of activity. *Acta Astronautica*, **45**(4–9), 449–56.

Roese, N.J., and Olson, J.M. (1995). Counterfactual thinking: a critical overview. In N.J. Roese and J.M. Olson (eds), *What Might Have Been: The Social Psychology Of Counterfactual Thinking*. Mahwah, NJ: Erlbaum.

Rogerson, W.P. (1989). Profit regulation of defense contractors and prizes for innovation. *The Journal of Political Economy*, **97**(6), 1284–1305.

Rogerson, W.P. (1994). Economic incentives and the defense procurement process. *Journal of Economic Perspectives*, **8**(4), 65–90.

Rosen, S. (1986). Prizes and incentives in elimination tournaments. *American Economic Review*, **76**(4), 701.

Ryan, R.M., and Deci, E.L. (2000). Self-determination theory and the facilitation of intrinsic motivation, social development, and well-being. *American Psychologist*, **55**(1), 68–78.

Saar, J. (2006). *Prizes: The Neglected Innovation Incentive*. Lund University.

Saldaña, J. (2009). *The Coding Manual for Qualitative Researchers*. Thousand Oaks: SAGE Publications Ltd.

Samuelson, W. (1986). Bidding for contracts. *Management Science*, **32**, 1533–50.

Sauser, B.J., Shenhar, A.J., and Hoffman, E.J. (2005). Identifying differences in space programs. In T.R. Anderson, T.U. Daim, D.F. Kocaoglu, D.Z. Milosevic, and C.M. Weber (eds), *Technology Management: A Unifying Discipline Technology Management: A Unifying Discipline*. Piscataway, NJ: IEEE Press, pp. 392–402.

Schroeder, A. (2004). *The Application and Administration of Inducement Prizes in Technology*. Golden, Colorado: Independence Institute.

Schrunk, D., Sharpe, B., Cooper, B., and Thangavelu, M. (2008). *The Moon. Resources, Future Development, and Settlement*. Berlin: Springer.

Scotchmer, S. (1999). On the optimality of the patent renewal system. *Rand Journal of Economics*, 30, 181–96.

Scotchmer, S. (2005). *Innovation and Incentives*. Cambridge, MA: MIT Press.

Seeni, A., Schafer, B., and Hirzinger, G. (2010). Robot mobility systems for planetary surface exploration – state of the art and future outlook: a literature survey. In T.T. Arif (ed.), *Aerospace Technologies Advancements*. Croatia: INTECH.

Shavell, S., and van Ypersele, T. (1999). Rewards versus intellectual property rights. National Bureau of Economic Research (NBER), Working Paper 6956.

Sidney, S. (1862). On the effect of prizes on manufactures. *Journal of the Society of Arts*, **25**(1), 1–52.

Sobel, D. (1996). *Longitude: The True Story of a Lone Genius Who Solved the Greatest Scientific Problem of His Time*. New York: Penguin.

SpaceRef.com. (2008). Rocket Racing Inc., Armadillo Aerospace and New Mexico Create Joint Venture to Launch Private Suborbital Space Transportation Business. Retrieved March 4, 2010, from http://www.spaceref.com/news/viewpr.html?pid=26813

SpaceX. (2011a). Space Exploration Technologies Corporation – Falcon 1. Retrieved February 2, 2011, from http://www.spacex.com/falcon1.php

SpaceX. (2011b). Space Exploration Technologies Corporation – Falcon 9. Retrieved February 2, 2011, from http://www.spacex.com/falcon9.php

Spear, A.J. (1995). Low cost approach to Mars Pathfinder and small landers. *Acta Astronautica*, **35**(Supplement 1), 345–54.

Steiner, C.J. (1995). A philosophy for innovation: the role of unconventional individuals in innovation success, *Journal of Product Innovation Management*. Blackwell Publishing Limited.

Stine, D.D. (2009). Federally funded innovation inducement prizes. Congressional Research Service.

Surrey Satellite Technology (SSTL). (2011). 'Smartphone Satellite' Developed by Surrey Space Researchers. Retrieved February 2, 2011, from http://www.sstl.co.uk/news-and-events?story=1706

Swink, M., Talluri, S., and Pandejpong, T. (2006). Faster, better, cheaper: a study of NPD project efficiency and performance tradeoffs. *Journal of Operations Management*, **24**(5), 542–62.

Tachikawa, K. (2007). Letter to Dr. Peter H. Diamandis, Chairman and Chief Executive Officer, X PRIZE Foundation. Retrieved February 2, 2011, from http://www.googlelunarxprize.org/files/downloads/lunar/jaxa_letter.pdf

Taylor, C.R. (1995). Digging for golden carrots: an analysis of research tournaments. *American Economic Review*, **85**(4), 872–90.

Team Phoenicia. (2011). How Are You Going To Fund Your GLXP Entry? Retrieved February 24, 2011, from http://www.googlelunarxprize.org/lunar/teams/team-phoenicia/blog/how-are-you-going-to-fund-your-glxp-entry

TrueZer0. (2008). TrueZer0. Retrieved June 20, 2010, from http://www.truezer0.com/

Vat, K.H. (2003). Toward an actionable framework of knowledge synthesis in the pursuit of learning organization. Paper presented at the Informing Science InSITE.

von Hippel, E. (1976). The dominant role of users in the scientific instrument innovation process. *Research Policy*, **5**(3), 212–39.

von Hippel, E. (1977). The dominant role of the user in semiconductor and electronic subassembly process innovation. *IEEE Transactions on Engineering Management*, EM-24(2), 60–71.

von Hippel, E. (1982). Appropriability of innovation benefit as a predictor of the source of innovation. *Research Policy*, **11**(2).

von Hippel, E. (1988). *The Sources of Innovation*. New York: Oxford University Press.

Vorder Bruegge, R.W. (1995). IAA International Conference on Low-Cost Planetary Missions April 12–15, 1994 Conference summary report. *Acta Astronautica*, **35**(Supplement 1), 771–8.

Waller, M.J., Conte, J.M., Gibson, C.B., and Carpenter, M.A. (2001). The effect of individual perceptions of deadlines on team performance. *Academy of Management Review*, **26**, 586–600.

Wei, M. (2007). *Should Prizes Replace Patents? A Critique of the Medical Innovation Prize Act of 2005*. SSRN.

Werner, D. (2010). 1-year deadline extension proposed for Google Lunar X Prize. Retrieved April 14, 2010, from http://spacenews.com/civil/100412-deadline-extension-google-lunar-prize.html

Whitaker, N. (2010). Interview with Norman Whitaker, DARPA Transformational Convergence Technology Office (TCTO), September 13, 2010.

White Label Space. (2010). PCB Design for Engine Throttle Controller. Retrieved February 20, 2011, from http://www.whitelabelspace.com/2010/05/pcb-design-for-engine-throttle.html

Williams, H. (2010). Incentives, prizes, and innovation. MIT and NBER (draft).

Wright, B.D. (1983). The economics of invention incentives: patents, prizes, and research contracts. *American Economic Review*, **73**(4), 691.

X Prize Foundation (XPF). (2004). Ansari X Prize. Retrieved February 4, 2010, from http://space.xprize.org/ansari-x-prize

X Prize Foundation (XPF). (2007). Armadillo Aerospace Nearly Wins Northrop Grumman Lunar Lander Challenge. Retrieved February 1, 2010, from http://www.xprize.org/llc/press-release/armadillo-aerospace-nearly-wins-northrop-grumman-lunar-lander-challenge

X Prize Foundation (XPF). (2008a). Google Lunar X Prize Q & A. Retrieved February 1, 2011, from http://www.googlelunarxprize.org/lunar/media-center/faq

X Prize Foundation (XPF). (2008b). Google Lunar X PRIZE Tech Talk: Will Pomerantz (Part 3 of 5). Retrieved February 2, 2011, from http://www.youtube.com/watch?v=SoPCyVkZrSM

X Prize Foundation (XPF). (2009). Google Lunar X PRIZE Panel at NewSpace 2009 (3 of 7). Retrieved February 2, 2011, from http://www.youtube.com/watch?v=ObjPANo30MI

X Prize Foundation (XPF). (2010). Team ARCA: From Ansari to the Google Lunar X Prize. Retrieved February 2, 2011, from http://www.youtube.com/watch?v=2WZfHkVc5gw

X Prize Foundation (XPF). (2011a). The Launch Pad: Was there ever something planned with the GLXP to do that was not done but would have significally altered it from what it now is? Did the GLXP become what it was intended to be? Retrieved February 18, 2011, from http://thelaunchpad.xprize.org/2011/02/was-there-ever-something-planned-with.html.

X Prize Foundation (XPF). (2011b). Prize Development. Retrieved February 2, 2011, from http://www.xprize.org/prize-development

Yin, R.K. (2003). *Case Study Research*. Thousand Oaks: SAGE Publications.

Zakrajsek, J.J., McKissock, D.B., Woytach, J.M., Zakrajsek, J.F., Oswald, F.B., McEntire, K.J., et al. (2005). *Exploration Rover Concepts and Development Challenges*. Cleveland, Ohio: NASA – Glenn Research Center.

Index